Geophysical Monograph Series

Including

IUGG Volumes
Maurice Ewing Volumes
Mineral Physics Volumes

Geophysical Monograph 133

Earth's Low-Latitude Boundary Layer

Patrick T. Newell
Terry Onsager
Editors

Ⓢ American Geophysical Union
Washington, DC

Published under the aegis of the AGU Books Board

Library of Congress Cataloging-in-Publication Data

Earth's low-latitude boundary layer / Patrick T. Newell, Terry Onsager, editors.
  p. cm. - (Geophysical monograph ; 133)
ISBN 0-87590-992-2
1.   Magnetospheric boundary layer. I. Newell, Patrick T., 1957- II. Onsager, Terry. III. Series.

QC809.M35 E17 2002
538'.766-dc21   2002034171

ISSN 0065-8448
ISBN 0-87590-992-2

CONTENTS

CONTENTS

CONTENTS

CONTENTS

Epilogue: Connections

PREFACE

We imagine the reader of this preface standing at the AGU bookstall wondering if the tome in hand is worth buying. The answer is "no", except for certain trifling exceptions. Those who wish to learn about the exciting pioneering years of LLBL research should buy the book for Tim Eastman's excellent historical review, our opening chapter. When did the term "LLBL" first enter the field? Eastman will tell you, and much else besides.

Graduate students, or experienced scientists changing their focus, will want to read Bengt Sonnerup's excellent tutorial and introduction to the modern understanding of the LLBL. The lead editor does not fit into either category, yet found Sonnerup's exposition enlightening. Sonnerup's overview, the longest single paper in the book, leads off our second chapter.

"The start is well enough," we imagine the preface reader (evidently a hardened cynic) grudgingly conceding. "But how does it end?" With a piece by George Siscoe and K. Siebert, which unites LLBL theory and observations with much else about magnetospheric interaction with the solar wind. Only those scientists interested in the unifying connections between merging, the LLBL, and the magnetospheric configuration will want to read it.

In the chapters between, leading scientists address such topics as the formation of the LLBL, the role of the LLBL as a plasma source and sink, and the relationship between the LLBL and the aurora, and other ionospheric phenomena. For example, Terry Onsager and Jack Scudder document Polar observations which appear to convincingly demonstrate that reverse merging poleward of the cusp for northward IMF can create both open and closed frontside LLBL field lines. This type of Song-Russell mechanism, which involves simultaneous uncorrelated merging in opposite hemispheres, was still considered quite controversial even a few years ago. The new Polar observations appear to have demonstrated that the Song-Russell mechanism does work, at least some of the time (see also the review by Paul Song in this volume).

However given the quasi-permanent nature of the LLBL, extending over huge volumes of space under a variety of IMF conditions, this single mechanism is unlikely to be the only means of LLBL formation, and may not be the major mechanism. Considerable evidence presented in this volume shows that wave-particle interactions can transport plasma across the magnetopause flanks, creating a closed boundary layer. Masaki Fujimoto's work on Geotail observations apparently demonstrates that such transport can occur, although it is unclear whether the observed surface waves observed are really Kelvin-Helmholtz. Katariina Nykyri, Antonius Otto, and colleagues explicate the theoretical issues involved.

The above and associated work raises the question of whether the LLBL serves as a major plasma source for the magnetotail. Those scientists interested in the source of magnetospheric plasma should also acquire this volume, as scientists with the most relevant observations cover the topic. For example, Marit Øieroset and colleagues presented Wind and Geotail observations showing that a dense cold plasma population appears along the magnetotail flanks, especially for northward IMF. This cold dense population is found within the plasma sheet, and covers an extended region. Øieroset also reports large standard deviations in velocity along the magnetotail flanks, consistent with turbulent mixing and entry mechanisms one might expect from the work just previously discussed.

A major theme of magnetospheric research over the last fifteen years has been the increasing ability to make quantitative inferences about magnetospheric dynamics from ionospheric observations, which are generally both cheaper and more comprehensive. Scientists interested in that trend will also benefit from this volume. A review by Sandholt and Farrugia introduces the burgeoning topic of how auroral observations now contribute substantially to our understanding of the boundary layers and their interactions with the solar wind. Likewise, Rodger and colleagues show how SuperDARN radar observations support the idea of antiparallel merging at distinct sites prenoon and postnoon in the opposite hemispheres.

Altogether then, only those researchers interested in the historical development of the field, or in its current theoretical state, or in the interaction of the boundary layers with the solar wind, or in how the boundary layers form, or in how solar wind plasma enters the magnetosphere, or in how ionospheric and auroral observations relate to the magnetospheric boundary layers, should purchase this book. We wish all others a happy retirement.

The editors gratefully acknowledge financial support from AFOSR, from the NSF Aeronomy program, and from NASA in producing this book. The original idea for an LLBL Chapman conference belonged to Walter Heikkila, who, along with Michael Lockwood, served ably as a convener. Ann Singer of AGU and Mary Ann Washington of JHU/APL were instrumental in the success of the meeting. Production of the book was possible only through the efforts of our AGU acquisition and production editors, Allan Graubard and Bethany Matsko, respectively.

Finally the following referees were instrumental in assuring that high scientific standards were maintained: B. Anderson, J. Borovsky, J. Chen, R. Clauer, S. Cowley, R. Denton, T. Eastman, M. Engebretson, C. Farrugia, M. Fujimoto, S. Fuselier, M. Hapgood, W. Heikkila, T. Horbury, J. Lemaire, K. Liou, R. Lysak, W. Lyatsky, N. Maynard, F. Menk, J. Moen, Z. Nemecek, P. Newell, K. Nykyri, M. Øieroset, T. Onsager, G. Parks, M. Prakash, A. Ridley, A. Rodger, M. Ruohoniemi, P. Sandholt, D. Sibeck, G. Siscoe, P. Song, Y. Song, T. Sotirelis, H. Stenuit, B. Sonnerup, K. Trattner, O. Troshichev, V. Vorobjev, R. Walker, S. Wing, R. Winglee, J. Woch, M. Yamaguchi, and X. Zhou.

P. T. Newell
The Johns Hopkins University
Applied Physics Laboratory

T. G. Onsager
NOAA Space Environment Center

Historical Review (pre-1980) of Magnetospheric Boundary Layers and the Low-Latitude Boundary Layer

Timothy E. Eastman

Institute for Science & Technology, Raytheon

INTRODUCTION

This historical survey covers research on the magnetospheric boundary layers through 1979 with a focus on observations and on the low-latitude boundary layer (LLBL) of Earth's magnetosphere. An effort was made to identify and read every paper published prior to 1980 that made a significant contribution towards identifying and characterizing boundary regions of the outer magnetosphere. The total list of over 80 papers is available at http://www.plasmas.org /BL. A sketch of Earth's magnetosphere is shown in Figure 1, which illustrates the major regimes and boundary layers. The magnetospheric boundary layer denotes all exterior boundary layers adjoining the magnetopause, including the dayside and tail flank portions of the LLBL, the exterior cusp region and entry layer, and the plasma mantle. There is a plasma sheet boundary layer separating the lobe and plasma sheet regions, but that boundary layer is not reviewed here [see *Eastman et al.*, 1984]. The magnetotail boundary layer refers to both the plasma mantle and the tail flank portion of the LLBL.

THEORY RELATED TO BOUNDARY LAYERS

Table 1 lists key theoretical developments prior to 1980. Shortly before Birkeland's auroral campaigns, Lord Kelvin offered "absolutely conclusive" evidence against any connection between the sun and magnetic storms [*Kelvin*, 1892]. In this context, it is all the more amazing that *Birkeland* [1896,1901,1908] first suggested a continuous solar wind and a cavity surrounding the Earth that excludes solar particles. Before the direct confirmation of the magnetopause from Explorer 12 [*Cahill and Amazeen*, 1963] and other in situ observations, which conclusively demonstrated the

continuous presence of both the solar wind and the magnetosphere, it was generally thought that both these systems and geomagnetic storms were intermittent phenomena. *Chapman and Ferraro* [1931] presumed such intermittence as they developed a model of the cavity boundary. *Ferraro* [1952] made this more quantitative and calculated the size, shape and thickness of the magnetopause.

Three major types of theories were proposed in the early 1960's for solar wind-magnetosphere interaction: dynamo models [*Piddington*, 1960], reconnection [*Dungey*, 1961], and viscous interaction [*Axford*, 1962]. A simple hydromagnetic flow model predicted solar wind plasma filling of the outer cusp region [*Spreiter and Summers*, 1967]. The increasingly successful reconnection model predicted cusp entry and convection into the magnetotail at high latitudes. Thus, by the early 1970s the focus was on observations in the cusp region and high latitudes, and *Frank* [1971] first clearly observed a cusp boundary layer using Imp 5 observations. In contrast, there was essentially no expectation at this time of finding a non-cusp boundary layer on the dayside at low latitudes. The viscous interaction and dynamo models remained without significant quantitative development with one notable exception. *Coleman* [1970, 1971] developed a comprehensive model for a dayside boundary layer using the dynamo concept, published in the Cosmic Electrodynamics and Radio Science journals, and which attracted little attention. Apparently unaware of these papers, *Willis* [1975] argued, on both theoretical and observational grounds, against any boundary layer other than the magnetopause layer for the dayside region at low latitude. He called attention to a clear crossing of the low-latitude magnetopause near local noon by Imp 5 presented by *Frank* [1971] who states that "the magnetopause near the magnetic equatorial plane appears to be an extremely effective barrier against the direct entry of solar plasma." In contrast, the theoretical expectation of boundary layer plasmas in the outer cusp region and tailward were clearly confirmed by a series of observations from 1971 onward as discussed in the next section.

Earth's Low-Latitude Boundary Layer
Geophysical Monograph 133
Copyright 2003 by the American Geophysical Union
10.1029/133GM01

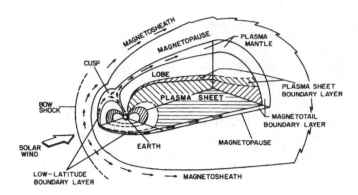

Figure 1. A sketch of the Earth's magnetosphere showing the boundary layers (adapted from *Eastman* [1979]).

Observations motivated four theoretical efforts in the late 1970's. One was the introduction by *Lemaire* [1977] of the impulsive penetration model, which emphasized the observed variability of solar wind input and analyzed the effect of filamentary current structures. Such filamentary structures were confirmed by *Russell and Elphic* [1978] who referred to them as "flux transfer events." The second was the evidence for high density gradients near the magnetopause based on high-resolution electron data [*Eastman and Hones*, 1979]. This was used by *Gary and Eastman* [1979] in an early quantitative study of the possible role of kinetic instabilities. Efforts to generate a boundary layer using MHD instabilities began to appear after 1980, although the Kelvin-Helmhotz instability was used in several earlier studies focused on the stability of the magnetopause [*Ong and Roderick*, 1972]. The third was the quantitative boundary layer to ionosphere coupling model developed by *Sonnerup* [1979] to test the suggestion by *Eastman et al.* [1976] that the LLBL acts as an MHD generator to drive field-aligned currents. The fourth key theoretical effort was that associated with magnetic merging or reconnection at the magnetopause and its associated energy dissipation. *Heikkila* [1975] clearly framed the problem and estimated the amount of energization expected, noting that available observations had so far failed to reveal the required energization. Indeed, finding such energization was a primary motivation for my analysis of Imp 6 plasma data near the magnetopause beginning late in 1975. Instead of finding the expected energization signatures, I instead confirmed the existence of the dayside LLBL and documented its primary signatures [see section below on the *Dayside Low-Latitude Boundary Layer*]. *Paschmann et al.* [1979] reported the first clear indication of plasma acceleration near the magnetopause as expected for magnetic merging.

The year of introduction for various terms relevant to magnetospheric boundaries is given in Table 2. Terminology for all major plasma regimes and boundaries was in place by the end of the 1970s.

CUSP AND PLASMA MANTLE OBSERVATIONS

Table 3 summarizes key observational results that pro-

vided indications and confirmation of the cusp, entry layer and plasma mantle regions. The cusp boundary layer was clearly identified by *Frank* [1971] using Imp 5 plasma observations combined with field model comparisons. Further confirmation was provided by *Russell et al.* [1971] and *Paschmann et al.* [1974].

In general, the direct confirmation of boundary layers near the magnetopause requires a combination of plasma and field observations. This is because magnetosheath and outer boundary layer spectra can be very similar, which creates the possibility that multiple magnetopause crossings with multiple magnetosheath segments can appear like a "boundary layer" to a plasma instrument. The Vela satellite series had no magnetometers because of their focus on the monitoring of atmospheric nuclear explosions. Nevertheless, *Hones et al.* [1972] showed possible boundary layer examples in Vela 4B plasma data that were primarily magnetosheath-lobe transitions and strongly suggested high-latitude boundary layer or plasma mantle. Working with Hones and the Los Alamos group, *Akasofu et al.* [1973] documented more cases with Vela 5 and 6, most of which were more than 5 R_E above the magnetotail neutral sheet. They interpreted this boundary

Table 1. Theory related to boundary layers

Year	Authors	Description
1896, 1901, 1908	Birkeland	First suggested solar wind and cavity around earth
1931	Chapman and Ferraro	First model of the magnetopause (MP)
1952	Ferraro	Calculated size, shape and thickness of the magnetopause
1957	Bierman	Estimated MP standoff distance
1961	Dungey	Reconnection model of solar wind - magnetosphere interaction
1964	Axford	Suggests complementary nature of reconnection, viscous interaction (*Axford*, [1962]), dynamo models (*Piddington*, 1960])
1964	Bernstein et al.	Production of broad MP transition from plasma instabilities
1967	Parker	Broadened MP from charge neutralization
1967	Spreiter and Summers	Predicted solar wind plasma filling of cusp region
1968	Eviatar and Wolf	Developed diffusive transfer theory for the magnetopause
1968	Stevenson and Comstock	Computer study of particle trajectories incident on magnetic field gradients
1970	Coleman	First comprehensive model for a low-latitude boundary layer; uses dynamo concept (*Piddington* [1960] and *Cole* [1960])
1975	Heikkila	Discusses problem of energy dissipation at the magnetopause
1975	Willis	Magnetopause review; argues against any boundary layer other than MP layer
1976	Lemaire	Impulsive penetration model; filamentary current structures
1979	Sonnerup	LLBL-ionosphere coupling model
1979	Gary and Eastman	Kinetic plasma model for boundary layer formation

layer as "the magnetic projection of the dayside cusps." An example of a magnetosheath-lobe transition with a boundary layer is shown in Figure 2a. The magnetopause is presumed to be associated with the first large density decrease during this inbound crossing. Temperature and bulk speed continue to decrease until the lobe is encountered sometime after 1800 UT. The latitude-longitude plot for the 18 R_E Vela sphere presented in Figure 2b shows that most cases occur at dZ > 5 R_E (or 16° latitude) and these crossings would be primarily transitions into the tail lobes as noted by the authors.

Haerendel [1974] reported the initial Heos 2 plasma mantle observations using both plasma and field observations. Confirmation of the plasma mantle was then completed by *Rosenbauer et al.* [1975] who provide a detailed report on Heos 2 observations. An example from their paper is shown in Figure 3, which shows a clear magnetopause near 1335 UT preceded by an extended boundary layer interval with intermediate density, speed, and temperature values. Evidence of ionospheric ions in a boundary layer were first reported by *Lundin et al.* [1979] based on Prognoz-7 observations in the plasma mantle.

Haerendel and Paschmann [1975] provided the first report and confirmation of the cusp-region entry layer, which has less turbulent flow signatures than the central cusp region. Figure 4 shows one example from their paper, which shows a clear magnetopause at 0657 UT. Unlike the dayside LLBL discussed below, they report a "lack of consistent flow direction" in the entry layer. Indeed, the example shown here has sunward flow convection in contrast to the anti-sunward flow component in the nearby magnetosheath. They also find the entry layer to be closely associated with the cusp region. *Haerendel and Paschmann* [1975] proposed that it is "causally related" in that magnetosheath plasma first enters the magnetosphere via the entry layer, followed by mirroring within the lower altitude cusp region, and then is convected upward and tailward to produce the plasma mantle. *Crooker* [1977] provided further confirmation of the entry layer using Explorer 33 observations.

Haerendel et al. [1978] discussed mantle, entry layer, and the LLBL observations with an emphasis on Heos 2 observations. The term "LLBL" was first coined in this paper.

TAIL FLANK BOUNDARY LAYER

Magnetotail boundary layer observations are summarized in Table 4. Cases specifically identified as magnetosheath-lobe transitions or plasma mantle are not included in this list. The flank tail boundary is noted by *Gosling et al.* [1967] as sometimes "not well defined." One or two possible boundary layer cases associated with magnetosheath-plasma sheet transitions are shown by *Hones et al.* [1972] although the lack of magnetometer data prevents clear magnetopause identification. The clearest example is shown in Figure 5 wherein a boundary layer is marked as an intermediate plasma region between the plasma sheet (high relative proton counts) and magnetosheath (highly spin modulated proton counts). *Akasofu et al.* [1973] show a similar case of proba-

ble flank boundary layer and refer to other cases in a dZ versus theta orbit plot. However, as noted above, most Vela cases were at high dZ and would have been primarily plasma mantle intervals.

Using a model combined with statistics of lunar orbit plasma data from Explorer 35 led *Howe and Siscoe* [1972] to infer a tail boundary layer with 2 R_E thickness. Similarly, *Hardy et al.* [1975] report flank boundary layer signatures at lunar distance. *Eastman et al.* [1976] state that "the boundary layer is nearly always present at all latitudes and longitudes at which Imp 6 crossed the magnetopause," which includes crossings back to about X = -6.6 R_E. Their focus and examples, however, were on dayside LLBL cases (see below).

Within a paper focused on magnetotail dynamics and plasma jetting, *Frank et al.* [1976] showed both plasma and field data for several plasma sheet to boundary layer transitions. The first interval in Figure 6a, prior to 0945 UT, appears to be a tail flank boundary layer although comparison magnetosheath plasma is not immediately adjoining this interval and a check of distribution functions would be needed to distinguish it from plasma sheet boundary layer. Within a paper by *Scarf et al.* [1977] focused on plasma

Table 2. Introduction of magnetospheric terminology

Year	Authors	Term or Concept
1946	Giovanelli	reconnection
1958	Parker	solar wind
1959	Gold	magnetosphere
1963	Sonett, Abrams	magnetopause
1964	Dessler	magnetosheath
1966	Bame et al.	plasma sheet
1972	Hones et al.	boundary layer
1975	Haerendel	plasma mantle
1978	Haerendel et al.	LLBL
1979	DeCoster, Frank	plasma sheet boundary layer

Table 3. Cusp and plasma mantle observations

Year	Authors	Observations
1971	Frank	Confirmation of cusp boundary layer; Imp 5 observations
1971	Russell et al.	Ogo 5 in high-altitude polar cusp
1972	Hones et al.	Indications of plasma mantle; MS-lobe transitions; Vela-no B
1973	Akasofu, Hones	~Hones; most cases dZ>5 Re; best two cases shown are magnetosheath to lobe
1974	Paschmann et al.	Polar cusp observations from Heos 2
1974	Haerendel	Initial plasma mantle observations
1975	Rosenbauer et al.	Detailed Heos 2 report on the plasma mantle; plasma and B data
1975	Haerendel, Paschmann	Confirmation of entry layer
1977	Crooker	Explorer 33 observations of entry layer
1978	Haerendel et al.	Review of Heos observations for mantle, entry layer, and LLBL

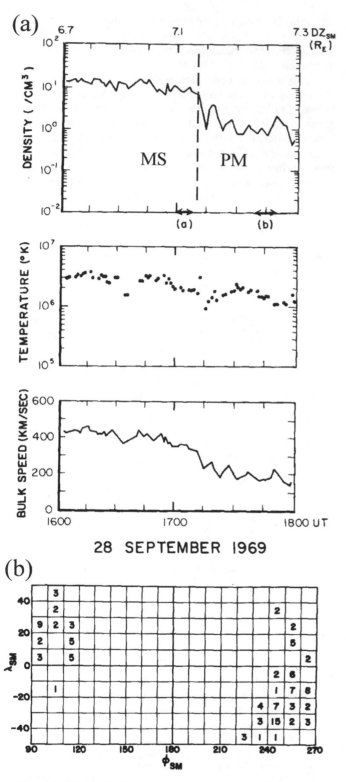

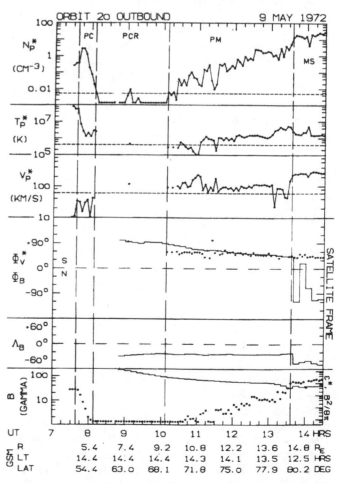

waves, there is a clear Imp 7 crossing of the flank LLBL shown here in Figure 6b and referred to by the authors as a "low-density mantlelike region." *Haerendel et al.* [1978] further documented characteristics of the flank LLBL and, as shown in Figure 6c, provided one example at 0535 local time. All together, these Imp 6, 7 and 8 and Heos 2 observations confirmed the presence and basic characteristics of the tail flank boundary layer.

Using Imp 6 plasma and field observations, *Eastman and Hones* [1979] reported statistics on 40 LLBL cases, including cases at X<0 of which one is shown in the paper. The most comprehensive study of the LLBL to date is my Ph.D. thesis, which provided detailed plasma and field data for 28 crossings, including four cases at X<0 [*Eastman, 1979*].

Dayside Low-Latitude Boundary Layer

The cusp and magnetotail boundary layers, either plasma mantle or flank type, are generally considered to be cusp-

Figure 2. First indications of plasma mantle. (a) Magnetosheath (MS) - lobe transition by Vela 5B with plasma mantle (PM) followed by lobe after 1800 UT (from *Akasofu et al,* [1973]) - Vela satellites had no magnetometer; (b) number of Vela 4B boundary crossings shown in 10x10 degree bins on the 18 R_E Vela sphere (from *Hones et al.,* [1972]).

Figure 3. Confirmation of plasma mantle. Lobe (marked PCR for polar cap region) to magnetosheath transition from Heos 2 with plasma mantle; magnetopause clearly indicated by magnetic field rotation near 1335 UT (from *Rosenbauer et al.* [1975]).

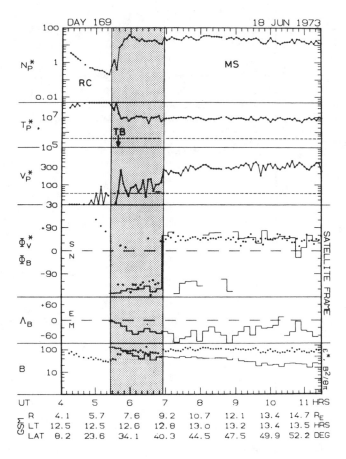

Figure 4. Confirmation of entry layer. Heos 2 transition from outer magnetosphere or ring current (RC) to magnetosheath with entry layer marked by shaded region (from *Haerendel and Paschmann* [1975]).

associated in this sense because plasma flows there are directed away from the subsolar region and not away from the cusp region [*Eastman and Hones,* 1979]. In contrast, flows can be either into or away from the cusp in the entry layer region. For example, the entry layer case in Figure 4 has sunward-directed plasma flow moving away from the cusp. As noted in the theory section above, early observations from Imp 5 presented by Frank [1971] indicated that the magnetopause at low latitude is "an extremely effective barrier." In addition to the arguments provided by *Willis* [1975] against any such boundary layer, many theorists were convinced that ideal MHD precluded direct entry of solar wind plasma across the magnetopause.

Table 5 summarizes dayside LLBL observations. As of the mid-1970s, there were no clear observations of the LLBL. The Vela results were all in the magnetotail, mostly sampled the plasma mantle, and lacked magnetometer data. Some papers in the mid-1960s refer to a "transition region" which, in context, is the magnetosheath. *Bonetti et al.* [1963] described some complex transition regions in the Explorer 10 data. The best early indication of a dayside boundary layer crossing is one unusual ATS 1 synchronous orbit magneto-

pause crossing reported by *Freeman et al.* [1968], which is shown in Figure 7a. Retarding potential analyzer (RPA) fluxes and flows are shown in two integral energy channels, E>0 eV and E>50 eV. A trade off is made between measuring flows and spectra, so spectra are unavailable. Some magnetosheath-like flows are shown near 0053 UT simultaneous with magnetospheric field levels based on high total field magnitudes. Later, *Ogilvie et al.* [1971] published examples of Ogo 5 electron and field data for some magnetopause crossings between 6.4 and 7.6 hours local time with a focus on transition intervals overlapping the current layer. As shown in Figure 7b, one Ogo 5 crossing shows a 50-second interval of elevated electron fluxes on magnetospheric field

Table 4. Magnetotail boundary layer observations*

Year	Authors	Spacecraft	Observations
1967	Gosling et al.	Vela 2	Outer tail boundary sometimes "not well defined"
1972	Hones et al.	Vela 4b	Indication of mantle and flank boundary layer (no B data)
1972	Howe, Siscoe	Exp 35	Tail boundary layer at lunar distance; model & statistics
1973	Akasofu et al.	Vela 5,6	Indication of mantle and flank BL; most cases dZ>5 Re
1975	Hardy et al.	Apollo	Indication of flank BL at lunar distance
1976	Frank et al.	Imp 8	Focus on plasma jetting; two examples of tail flank BL
1976	Eastman et al.	Imp 6	Frontside LLBL focus; tail flank LLBL cases noted
1977	Scarf et al.	Imp 7	Focus on plasma waves; one example of tail flank BL
1978	Haerendel et al.	Heos 2	Heos summary of plasma mantle, entry layer and LLBL
1979	Eastman, Hones	Imp 6	Statistics for 40 LLBL cases; one example shown at X<0
1979	Eastman	Imp 6	Ph.D. thesis on LLBL; details for 28 crossings; 4 at X<0

*Table does not include papers specifically focused on "plasma mantle."

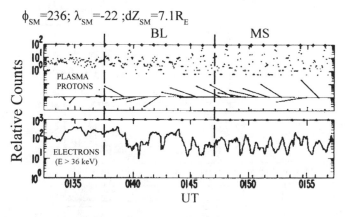

Figure 5. First indication of tail flank boundary layer. Vela transition from plasma sheet (PS), through possible boundary layer, to magnetosheath (from *Hones et al.,* [1972]).

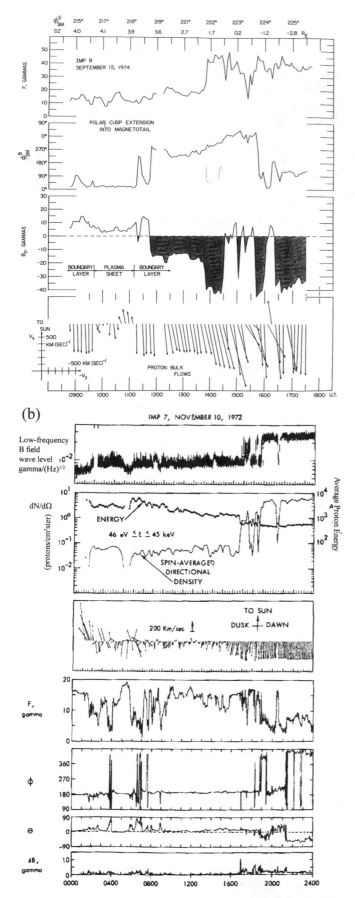

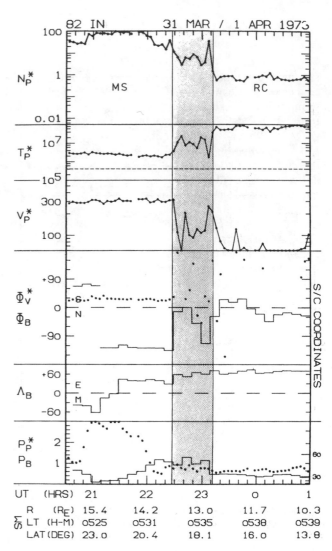

Figure 6. Confirmation of tail flank boundary layer. (a) Possible tail flank boundary layer from Imp 8 shown prior to 0945 UT (from *Frank et al.* [1976]); (b) Clear tail flank LLBL from Imp 7 showing both plasma and field data (from *Scarf et al.* [1977]); (c) Heos sample of tail flank LLBL (from *Haerendel et al* [1978]).

lines (not noted by the authors). Although ion data and flows were not presented, this interval suggests the presence of a boundary layer distinct from the magnetopause layer. Convection electric fields were measured by Aggson in a thin layer adjacent to the magnetopause but plasma densities were not reported [see *Heppner*, 1972]. In a review paper presented at the International Symposium on Solar-Terrestrial Physics in Boulder (June 7-18), *Sonnerup* [1976] included a section entitled "Front Lobe Boundary Layer." He states that "While the existence and properties of the high latitude plasma mantle are well documented, little direct experimental information exists concerning the presence or absence on the front lobe of the magnetosphere of a similar plasma boundary layer adjacent to the magnetopause."

Simultaneous with Sonnerup's review presentation, AGU received our manuscript, which provided details on combined ion and magnetic field data for two dayside crossings shown here in Figure 8 [*Eastman et al.*,1976; see Appendix for personal reflections]. This research with the Los Alamos group was based on an extensive study of Imp 6 ion, electron, and field data for the low-latitude boundary region from the tail flank at X = -6.6 R_E to the subsolar region. Although full documentation for 28 LLBL crossings was delayed for the thesis itself [*Eastman*, 1979], this paper confirmed the existence of the dayside LLBL. The magnetopause location and thickness is unambiguous for the February 4, 1972 LLBL case. Detailed magnetic field plots show the current layer interval to be at 0050 to 0052:30 UT. Magnetosheath-like flows and intermediate density and ion energies clearly mark the boundary layer interval. The magnetopause interval is much less distinct in the June 16 case but extends no later than about 0406 UT based on full field and plasma characteristics [see Figure 2.3.4.2 in *Eastman*, 1979]. This crossing represents an unusually smooth boundary layer transition. Electron density samples at 12.5-s time interval suggested substantial sub-structure within the LLBL [*Eastman and Hones*, 1979], which led *Gary and Eastman* [1979] to propose a possible role for drift instabilities, especially the lower-hybrid drift instability.

Only one of the 28 LLBL crossings shown in *Eastman* [1979] has a possible density plateau signature as reported for some Heos 2 crossings by *Haerendel et al.* [1978]. Apparent plateaus can result from instrument cycle times, which were 256 seconds for Heos 2 compared with 100 seconds for the Imp 6 ion data. Clear density plateau cases were later published based on ISEE observations, but a plateau signature was not common for boundary layer crossings reported prior to 1980.

SUMMARY

Observational papers demonstrating the existence of major boundary regions of Earth's magnetosphere are listed in Table 6. *Sonett* [1959] reported an early observation of a probable magnetopause crossing from Pioneer 1 and 5. However, the presence of a magnetopause is inferred from patchy field samples. First confirmation or completion of the discovery process was accomplished with the following papers: *Cahill and Amazeen* [1963] for the magnetopause; *Frank* [1971] for the cusp boundary layer; *Haerendel* (1974) and *Rosenbauer et al.* [1975] for the plasma mantle; *Haerendel and Paschmann* [1975] for the entry layer; and *Eastman et al.* [1976] for the dayside LLBL.

The discovery process is often not based on any single observation or even any single publication. Confirmation of a boundary layer crossing requires a combination of plasma and field measurements. Magnetic field measurements are usually required to uniquely identify the magnetopause or current layer interval. Multiple magnetopause crossings could give the signatures of a boundary layer when sampling with a low-resolution plasma instrument.

For many researchers, *Hones et al.* [1972] "discovered" the tail flank LLBL. As shown above and specifically stated

by Akasofu et al. [1973], most of the probable Vela boundary layer crossings occurred at high dZ values in association with magnetosheath-lobe transitions. Thus, if Hones discovered the tail flank LLBL, then he discovered the plasma mantle as well. However, discovery of the plasma mantle has long been credited to the Heos 2 group. Further, the Vela satellite had no magnetometer and unique magnetopause identification was not possible. Boundary layer intervals distinct from the magnetopause layer could be inferred or suggested, but could not be confirmed.

In my opinion, it is appropriate to credit *Hones et al.* [1972] with first indications of both the plasma mantle and tail flank boundary layers - in their own words, the "magnetotail boundary layer." Confirmation then followed with the Heos 2 results of *Haerendel* (1974) and *Rosenbauer et al.* [1975]. However, first confirmation or completion of the discovery process is more complicated with the tail flank LLBL. *Eastman et al.* [1976] noted that the boundary layer is "nearly always present at all" local times, which for Imp 6 means crossings back to about X = -6.6 R_E. A later study of magnetopause crossings without a boundary layer indicate that "such crossings constitute about 10% of all magnetopause crossings" [*Eastman*, 1996]. Although confirming the dayside LLBL, *Eastman et al.* [1976] did not include any tail flank LLBL crossings. Such cases were published in our 1979 papers [*Eastman and Hones*, 1979; *Eastman*, 1979]. *Frank et al.* [1976] showed a possible tail flank boundary layer crossing without identification as such. A clear tail flank LLBL case with both plasma and field data was later published by Scarf et al. [1977]. *Haerendel et al.* [1978] further documented characteristics of the flank LLBL with combined plasma and field signatures. In combination, observations from Imp 6, 7, 8, and Heos 2 confirmed the existence of the tail flank LLBL and documented its basic characteristics.

Table 5. Dayside LLBL observations

Year	Authors	Spacecraft	Observations
1963	Bonetti et al.	Exp 10	Associated complex transitions with possible BL (MS ?)
1966	Wolfe et al.	Imp 1	Some passes with gradual ion flux decrease near MP
1968	Freeman et al.	ATS 1	Indication of LLBL with one event at geostationary orbit
1971	Ogilvie et al.	Ogo 5	Possible LLBL in electron data near dawn meridian
1976	Eastman et al.	Imp 6	Confirmation of dayside LLBL; two examples
1978	Haerendel et al.	Heos 2	Heos summary; six examples; first used "LLBL"
1979	Eastman, Hones	Imp 6	Statistics for 40 LLBL cases; 11 examples presented
1979	Eastman	Imp 6	Ph.D. thesis on LLBL; details for 28 crossings
1979	Paschmann et al.	ISEE	First indication of merging acceleration at magnetopause

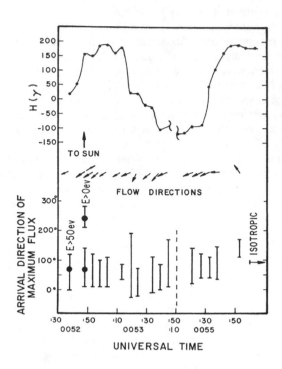
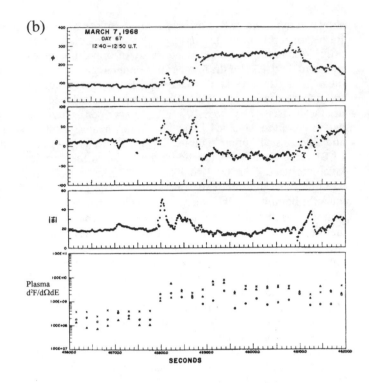

Figure 7. First indications of dayside LLBL. (a) ATS 1 synchronous orbit crossing at 14 hours local time showing total magnetic field H and RPA fluxes and flows for E>0 eV and E>50 eV (from *Freeman et al.* [1968]); (b) Possible brief LLBL period in Ogo 5 electron data (from *Ogilvie et al* [1971]).

CONCLUSION

The period from the advent of the space age in the late 1950s up through the 1970s was a great period of discovery for magnetospheric physics in which most of its major regimes and boundaries were first identified and characterized. This review has attempted to encompass all publications that enabled these discoveries with a focus on major plasma regions and boundaries of the Earth's outer magnetosphere. Magnetospheric physics has continued to be a major area of space physics studies, most recently with a focus on "Space Weather" and space environment impacts on technological systems. However, the discovery mode so dominant prior to 1980 is now replaced by a much greater balance of theory, modeling, laboratory experiment, and in situ observation that promises continuing progress in quantitative understanding of one of the most complex, large-scale physical systems in science.

APPENDIX. PERSONAL REFLECTIONS

Working with Ed Hones and the Los Alamos group, I began in mid-1975 to carry out a systematic analysis of Vela and Imp 6 magnetopause crossings. Dr. Hones was busy working on magnetotail and substorm issues and graciously allowed me to focus my thesis work, with the Geophysical Institute of the University of Alaska, on the boundary layer question, which he had initiated earlier with the Vela studies.

The Ph.D. thesis topic proposal of July 23, 1975 refers to "A Study of the Magnetotail Boundary Layer." For the Fall 1975 AGU meeting, my presentation was on "Recent Vela satellite observations of the magnetospheric boundary layer." After that presentation (results unpublished), I turned my attention entirely over to the Imp 6 data set with Hones' encouragement and our mutual recognition of the importance of simultaneous plasma and magnetic field measurements in the dayside magnetopause region. My initial analysis encompassed some 225 Imp 6 magnetopause crossings which I quickly pared down to about 133 cases for thickness estimates and a search for any correlations with solar wind input and geomagnetic activity. Ultimately, detailed analysis focused on 28 LLBL crossings for the thesis [*Eastman*,1979]. Early on, my objective was to look for evidence of accelerated flows near the magnetopause on the dayside or other evidence of energization expected for merging. Based on a comprehensive search of Imp 6 magnetopause crossings at all local times, and using combined ion, electron, and magnetic field data, I instead found clear evidence in many crossings for a distinct boundary layer located earthward of the current layer. Recognizing the importance of these results, I worked rapidly to prepare a paper for the newly formed journal, Geophysical Research Letters. My first draft, dated April 1, 1976, used the term "transfer layer" for the dayside boundary layer. In the final published work, we used the term "magnetospheric boundary layer" [*Eastman et al.,* 1976].

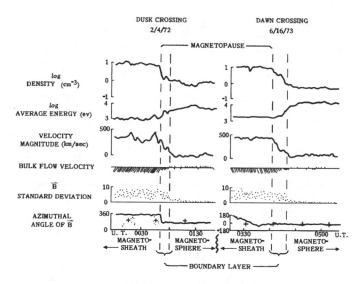

Figure 8. Confirmation of dayside LLBL. Duskside and dawnside LLBL crossings by Imp 6 (from *Eastman et al.* [1976]).

Table 6. Magnetopause and boundary layer observations

Year	Authors	Spacecraft	Observations
1963	Cahill, Amazeen	Exp 12	Confirmation of magnetopause
1968	Freeman et al.	ATS 1	First indication of dayside LLBL
1971	Frank	Imp 5	Confirmation of cusp boundary layer
1972	Hones et al.	Vela 4b	First indication of plasma mantle and tail flank boundary layer
1974	Haerendel	Heos 2	Initial report of plasma mantle
1975	Rosenbauer et al.	Heos 2	Confirmation of plasma mantle
1975	Haerendel and Paschmann	Heos 2	Confirmation of entry layer
1976	Frank et al.	Imp 8	Possible tail flank case
1976	Eastman et al.	Imp 6	Confirmation of dayside LLBL; noted tail flank cases
1977	Scarf et al.	Imp 7	First clear flank LLBL case with both plasma and field data
1978	Haerendel et al.	Heos 2	Heos summary; further confirmation of tail flank and dayside LLBL
1979	Eastman, Hones	Imp 6	Statistics for 40 LLBL cases; both dayside and tail flank crossings

Karl Schindler was visiting Los Alamos at about that time. I recall in particular how skeptical Dr. Schindler was about my evidence for a boundary layer distinct from and earthward of the current layer. We noted how energetic electron data indicated, at least for some intervals, that the boundary layer is on closed field lines. Dr. Schindler understood that ideal MHD would not allow for the transfer of magnetosheath plasma across the current layer from a region of open to closed field lines. He asked us whether what we are observing was really just an extended current layer or whether there might be a problem with the data. I spent many additional hours checking for any possibility of data errors, such as a timing error between the magnetic field and plasma

data. These checks were helpful in insuring the validity of our results as we prepared for publications in 1976 and later.

The title of our 1976 paper was "The Magnetospheric Boundary Layer: Site of Plasma, Momentum and Energy Transfer from the Magnetosheath into the Magnetosphere." In addition to documenting the presence and characteristics of the LLBL, we presented qualitative model and observational tests for such transfer by treating the boundary layer as an MHD generator. The earlier theoretical study by *Coleman* [1970, 1971] had many similar features as an MHD generator model but Coleman's work was unknown to us before the final draft. In addition to various specifics on the generation and mapping of field-aligned currents into the ionosphere, we emphasized the "longitudinal spreading of depolarizing currents" due to large non-dipolar field distortions near the outer magnetospheric boundary. As I reported in a paper presented at the Second Magnetospheric Cleft Symposium in St. Jovite, Quebec in October 5-8, 1976, one consequence of this mapping is that a set of localized boundary interactions on the dayside outer boundary would map down to the ionospheric cusp region as a set of "cat whisker" features. In that presentation, I pointed out the striking similarity of our prediction with optical data from DMSP [*Snyder and Akasofu*, 1976]. Later modeling and dayside cusp observations confirmed this early suggestion of a close association between cusp auroral features and the dayside LLBL [*Crooker and Siscoe*, 1990].

Acknowledgments. I am grateful to the many scientists that encouraged my work in the field of space plasma physics. For the period relevant to this review, I am especially thankful for the inspiration and support provided by Edward Hones, Jr., S.-I. Akasofu, S. Peter Gary, Karl Schindler, and Walter Heikkila.

REFERENCES

Akasofu, S.-I., E. W. Hones, Jr., S. J. Bame, J. R. Asbridge, and A. T. Y. Lui, Magnetotail and boundary layer plasmas at a geocentric distance of ~18 R_E: Vela 5 and 6 observations, *J. Geophys. Res., 78,* 7257-7274, 1973.

Axford, W. I., The interaction between the solar wind and the Earth's magnetosphere, *J. Geophys. Res., 10,* 3791-3796, 1962.

Axford, W. I., Viscous interaction between the solar wind and the Earth's magnetosphere, *Planet. Space Sci., 12,* 45-54, 1964.

Bame, S. J., J. R. Asbridge, H. E. Felthauser, R. A. Olson, and I. B. Strong, Electrons in the plasma sheet of the Earth's magnetic tail, *Phys. Rev. Lett, 16,* 138, 1966.

Bernstein, W., R. W. Fredericks, and F. L. Scarf, A model for a broad disordered transition between the solar wind and the magnetosphere, *J. Geophys. Res., 69,* 1201-1210, 1964.

Biermann, L., Solar corpuscular radiation and the interplanetary gas, *Observatory, 77,* 109-110, 1957.

Birkeland, Kr., Sur les rayons cathodiques sons l'action de forces magnetiques intenses, *Arch. Sci. Phys. Naturelles, 1,* 497-512 (and plate VII facing p. 592), 1896.

Birkeland, Kr., Expedition norvegienne de 1899-1900 pour l'etude des aurores boreales. Resultats des recherches magnetiques, Videnskap. Skrifter, I, Math.-Naturv. K1, 1, 1901.

Birkeland, Kr., *The Norwegian Aurora Polaris Expedition 1902-3, vol. 1, On the Cause of Magnetic Storms and the Origin of Terrestrial Magnetism, first section,* H. Aschehoug and Co., Christiania, 1908.

Birkeland, Kr., *The Norwegian Aurora Polaris Expedition 1902-3, vol. 1, On the Cause of Magnetic Storms and the Origin of Terrestrial Magnetism, second section*, H. Aschehoug and Co., Christiania, 1913.

Bonetti, A., H. S. Bridge, A. J. Lazarus, B. Rossi, and F. Scherb, Explorer 10 plasma measurements, *J. Geophys. Res., 68,* 4017-4063, 1963.

Cahill, L. J., and P. G. Amazeen, The boundary of the geomagnetic field, *J. Geophys. Res., 68,* 1835-1843, 1963.

Chapman, S., and V. C. A. Ferraro, A new theory of magnetic storms: Part I - The initial phase, *Terr. Magn. Atmos. Electr., 36,* 77-97, 171-186, 1931.

Cole, K. D., A dynamo theory of the aurora and magnetic disturbance, Aust. J. Phys., 13, 484-497, 1960.

Coleman, P. J., Jr., Tangential drag on the geomagnetic cavity, *Cosmic Electrodyn. 1,* 145-159, 1970.

Coleman, P. J., Jr., A model of the geomagnetic cavity, *Radio Science, 6,* 321-340, 1971.

Crooker, N. U., Explorer 33 entry layer observations, *J. Geophys. Res., 82,* 515-522, 1977.

Crooker, N. U., and G. L. Siscoe, On mapping flux transfer events to the ionosphere, *J. Geophys. Res., 95,* 3795-3799, 1990.

DeCoster, R. J., and L. A. Frank, Observations pertaining to the dynamics of the plasma sheet, 84, *J. Geophys. Res., 84,* 5099,1979.

Dessler, A. J., Length of the magnetosphere tail, *J. Geophys. Res., 69,* 3913, 1964.

Dungey, J. W., Interplanetary magnetic field and the auroral zones, *Phys. Rev. Lett., 6,* 47, 1961.

Eastman, T. E., E. W. Hones, Jr., S. J. Bame, and J. R. Asbridge, The magnetospheric boundary layer: Site of plasma, momentum and energy transfer from the magnetosheath into the magnetosphere, *Geophys. Res. Lett., 3,* 685-688, 1976.

Eastman, T. E., and E. W. Hones, Jr., Characteristics of the magnetospheric boundary layer and magnetopause layer as observed by Imp 6, *J. Geophys. Res., 84,* 2019-2028, 1979.

Eastman, T. E., *The Plasma Boundary Layer and Magnetopause Layer of the Earth's Magnetosphere,* Ph.D. Thesis, University of Alaska, 1979 [Los Alamos Scientific Laboratory Report LA-7842-T, June, 1979]

Eastman, T. E., W. K. Peterson, W. Lennartsson, and L. A. Frank, The plasma sheet boundary layer, *J. Geophys. Res., 89,* 1553-1572, 1984.

Eviatar, A. and R. A. Wolf, Transfer processes in the magnetopause, J. Geophys. Res., 73, 5561-5576, 1968.

Ferraro, V. C. A., On the theory of the first phase of a geomagnetic storm: A new illustrative calculation based on an idealized (plane not cylindrical) model field distribution, *J. Geophys. Res., 57,* 15-49, 1952.

Frank, L. A., Plasma in the Earth's polar magnetosphere, *J. Geophys. Res., 76,* 5202-5219, 1971.

Frank, L. A., K. L. Ackerson, and R. P. Lepping, On hot tenuous plasmas, fireballs, and boundary layers in the Earth's magnetotail, *J. Geophys. Res., 81,* 5859-5881, 1976.

Freeman, J. W., Jr., C. S. Warren, and J. J. Maguire, Plasma flow directions at the magnetopause on January 13 and 14, 1967, *J. Geophys. Res., 73,* 5719-5731, 1968.

Gary, S. P., and T. E. Eastman, The lower hybrid drift instability at the magnetopause, *J. Geophys. Res., 84,* 7378-7381, 1979.

Giovanelli, R. G., A theory of chromospheric flares, *Nature, 158,* 81, 1946.

Gold, T., Motion in the magnetosphere, *J. Geophys. Res., 64,* 1219, 1959.

Gosling, J. T., J. R. Asbridge, S. J. Bame, and I. B. Strong, Vela 2 measurements of the magnetopause and bow shock positions, *J. Geophys. Res., 72,* 101-112, 1967.

Haerendel, G., Die Spur der Magnetopause in der Magneto-sphare, *Mitt. Astron. Ges., 35,* 165, 1974.

Haerendel, G., and G. Paschmann, Entry of solar wind plasma into the magnetosphere, in *Physics of the Hot Plasma in the Magnetosphere,* B. Hultqvist and L. Stenflo, eds., Plenum Press, N.Y., pp. 23-43, 1975.

Haerendel, G., G. Paschmann, N. Sckopke, H. Rosenbauer, and P. C. Hedgecock, The frontside boundary layer of the magnetosphere and the problem of reconnection, *J. Geophys. Res., 83,* 3195-3216, 1978.

Hardy, D. A., H. K. Hills, and J. W. Freeman, A new plasma regime in the distant geomagnetic tail, *Geophys. Res. Lett., 2,* 169-172, 1975.

Heikkila, W. J., Is there an electrostatic field tangential to the dayside magnetopause and neutral line?, *Geophys. Res. Lett., 2,* 154-157, 1975.

Heppner, J. P. Electric fields in the magnetosphere, in *Critical Problems of Magnetospheric Physics,* E. R. Dyer, ed., National Academy of Sciences, Washington, D.C., 107-120, 1972.

Hones, E. W., Jr., J. R. Asbridge, S. J. Bame, M. D. Montgomery, S. Singer, and S.-I. Akasofu, Measurements of magnetotail plasma flow made with Vela 4B, *J. Geophys. Res., 77,* 5503-5522, 1972.

Howe, H. C., and G. L. Siscoe, Magnetopause motions at lunar distance determined from the Explorer 35 plasma experiment, *J. Geophys. Res., 77,* 6071-6086, 1972.

Kelvin, W. T., Address to the Royal Society at their Anniversary Meeting, Nov. 30, 1892, *Proc. Roy. Soc. London, A, 52,* 300-310, 1892.

Lemaire, J., Impulsive penetration of filamentary plasma elements into the magnetospheres of the Earth and Jupiter, *Planet. Space Sci., 25,* 887-890, 1977.

Lundin, R., et al., First observations of the hot ion composition in the high latitude magnetospheric boundary layer by means of Prognoz-7, in *Proceedings of Magnetospheric Boundary Layers Conference,* Alpbach, 11-15 June 1979 (ESA SP-148), pp. 91-96, August, 1979.

Ogilvie, K. W., J. D. Scudder, and M. Sugiura, Magnetic field and electron observations near the dawn magnetopause, *J. Geophys. Res., 76,* 3574-3586, 1971.

Ong, R. S. B., and N. Roderick, On the Kelvin-Helmholtz instability of the Earth's magnetopause, *Planet. Space Sci., 20,* 1-10, 1972.

Parker, E. N., Interaction of the solar wind with the geomagnetic field, *Phys. Fluids, 1,* 171-187, 1958.

Parker, E. N., Small-scale nonequilibrium of the magnetopause and its consequences, *J. Geophys. Res., 72,* 4365-4374, 1967.

Paschmann, G., H. Grunwaldt, M. D. Montgomery, H. Rosenbauer, and N. Sckopke, Plasma observations in the high-latitude magnetosphere, in *Correlated Interplanetary and Magnetospheric Observations,* D. E. Page, ed., D. Reidel, Dordrecht, Holland, pp. 249-253, 1974.

Paschmann, G., B. U. O. Sonnerup, I. Papamastorakis, N. Sckopke, G. Haerendel, S. J. Bame, J. R. Asbridge, J. T. Gosling, C. T. Russell, and R. C. Elphic, Plasma acceleration at the Earth's magnetopause: Evidence for reconnection, *Nature, 282,* 243-246, 1979.

Piddington, J. H., A theory of polar geomagnetic storms, *Geophys. J. Roy. Astr. Soc., 3,* 314-332, 1960.

Rosenbauer, H., H. Grunwaldt, M. D. Montgomery, G. Paschmann, and N. Sckopke, Heos 2 plasma observations in the distant polar magnetosphere: The plasma mantle, *J. Geophys. Res., 80,* 2723-2737, 1975.

Russell, C. T., C. R. Chappell, M. D. Montgomery, M. Neugebauer, and F. L. Scarf, Ogo 5 observations of the polar cusp on November 1, 1968, *J. Geophys. Res., 76,* 6743-6764, 1971.

Russell, C. T., and R. C. Elphic, Initial ISEE magnetometer results: magnetopause observations, *Space Sci. Rev., 22,* 681-715, 1978.

Scarf, F. L., L. A. Frank, and R. P. Lepping, Magnetospheric boundary observations along the Imp 7 orbit 1. Boundary locations and wave level variations, *J. Geophys. Res., 82,* 5171-5180, 1977.

Snyder, A. L., Jr., and S.-I. Akasofu, Auroral oval photographs from the DMSP 8531 and 10533 satellites, *J. Geophys. Res., 81,* 1799-1804, 1976.

Sonett, C. P., Magnetic compression in the geomagnetic field as measured by Pioneer I, *Astro. Soc. Pac., 71,* 369, 1959.

Sonett, C. P., and I. J. Abrams, The distant geomagnetic field, 3. Disorder and shocks in the magnetosphere, *J. Geophys. Res., 68,* 1233, 1963.

Sonnerup, B. U. O., Magnetopause and boundary layer, in *Physics of Solar Planetary Environments,* D. J. Williams, ed., American Geophysical Union, Washington, D.C. 541-557, 1976.

Sonnerup, B. U. O., Theory of the low latitude boundary layer, in *Proceedings of Magnetospheric Boundary Layers Conference,* Alpbach, 11-15 June 1979 (ESA SP-148), pp. 395-397, August, 1979.

Spreiter, J. R., and A. L. Summers, On conditions near the neutral points on the magnetospheric boundary, *Planet. Space Sci., 15,* 787, 1967.

Stevenson, T. E., and C. Comstock, Particles incident on magnetic field gradients, *J. Geophys. Res., 73,* 175-184, 1968.

Willis, D. M., The microstructure of the magnetopause, *Geophys. J. R. astro. Soc., 41,* 355-389, 1975.

Wolfe, J. H., R. W. Silva, and M. A. Meyers, Observations of the solar wind during the flight of IMP 1, *J. Geophys. Res., 71,* 1319-1340, 1966.

T. E. Eastman, *Institute for Science & Technology, Raytheon,* Code 630, NASA Goddard Space Flight Center, Greenbelt, MD 20771 (E-mail: timothy.eastman @gsfc.nasa.gov); also Consultant with Plasmas International, 1225 Edgevale Road, Silver Spring, MD 20910. (E-mail: plasmas@verizon.net)

Theory of the Low Latitude Boundary Layer and its Coupling to the Ionosphere: A Tutorial Review

Bengt U. Ö. Sonnerup

Thayer School of Engineering, Dartmouth College, Hanover, NH

Keith D. Siebert

Mission Research Corporation, Nashua, NH

A tutorial overview is given of theoretical models that describe the plasma flow across closed field lines in the low latitude boundary layer and the perfect or imperfect coupling of the layer to the dayside auroral ionosphere by means of field-aligned currents. Forces that control the flow include $\mathbf{j} \times \mathbf{B}$ forces, viscous forces, pressure forces, and inertia forces. Characteristic scale sizes for the boundary layer width are derived and important dimensionless groups are identified. A comparison of model results is made with results from a global numerical simulation, using the ISM code for the case of vanishing interplanetary magnetic field. Application of the theory when most of the boundary layer is on open field lines is discussed.

1. INTRODUCTION

As its title indicates, this article is concerned with theoretical aspects of the coupling between the low latitude boundary layer (the LLBL) and the high-latitude ionosphere. This means that observational facts are presented only when they are directly relevant to some prominent feature of the theoretical models. The paper is tutorial in character and therefore reviews many aspects of the theory that are now old and well known to the experts but that students and others entering the field may find useful. Also, for tutorial purposes the models have been reduced to their simplest and most fundamental ingredients. For example, the ionosphere is treated as a conducting layer, mostly with constant conductivity. An important objective of the paper is to derive expresqsions for the spatial scale sizes and the dimensionless groups that characterize the LLBL.

The paper is a review in the sense that it contains no new original developments but only some new insights and generalizations of old models. No attempt is made to review all of the individual contributions that have led to the understanding we now have of the theory. Therefore the reference list is not as comprehensive as, strictly speaking, it should be. The reader may find it useful to refer to the more complete review by *Lotko and Sonnerup* [1995] and to the articles cited in the original papers upon which our presentation is based.

Three comments should be made at the outset. First, the LLBL of interest to us contains plasma that is moving across field lines away from the subsolar region, although return flow phenomena are included as well. The origin of the LLBL plasma is, for the most part,

Earth's Low-Latitude Boundary Layer
Geophysical Monograph 133
10.1029/133GM02

ary conditions so permit, (in particular, one needs $dV_0/dx = 0$), these solutions will be asymptotically reached at large distances downstream. But in reality there will probably always be significant variations in the x direction, i.e., $\partial/\partial x \neq 0$. As long as turbulent fluctuations are averaged over, the changes along x are expected to be much more gradual than those along y so that $\partial/\partial x << \partial/\partial y$. In this situation, the boundary-layer approximation, well known from ordinary fluid mechanics [e.g., *Schlichting*, 1968], is applicable and has been used by *Drakou et al.* [1994], along with a marching numerical integration scheme appropriate for the parabolic (viscous) momentum equation, to obtain solutions for constant H that include compressibility effects as well as an approximate description of the self-consistent magnetic field in the LLBL in terms of a power series expansion in the z coordinate. The model contains inertia forces and pressure forces, in addition to viscous forces and $\mathbf{j} \times \mathbf{B}$ forces but it does not include $\Delta\Phi_\parallel$ in the coupling region. For a reasonably realistic set of boundary conditions, one observes gradual entrainment of the magnetospheric plasma into the tailward flow of the LLBL plasma and also a local maximum in the field-aligned current density at a flow distance of some $7.5R_E$ from the subsolar point, i.e., around 9AM and 3PM local time.

Wei et al. [1996] developed a numerical model similar to that of *Drakou et al.* but they also included field-aligned potential drops, given by (33), as well as mapping factors dx/dx_i and dy/dy_i that are realistic functions of both x and y. They compared simulation results for different levels of the mapping factors, using $\kappa = 10^{-9}$ mho/m^2 as well as $\kappa = 0$, with results from the Viking spacecraft for the latitudinal width of the ionospheric footprint of the LLBL [*Woch et al.*, 1993]. The conclusion reached was that the numerical results could be reasonable fitted to the observations for $dx/dx_i \simeq 60$ and $\kappa \simeq 10^{-9}$ mho/m^2.

Sample results from the *Wei et al.* simulation are shown in Figure 6. The two top left panels show the viscous entrainment of hot tenuous magnetospheric plasma into the tailward (downward in the figure) flow. The top right panel shows a maximum in Region 1 current density at a flow distance $x = 15 - 20R_E$. The remaining panels in the figure show profiles at three different x values of velocity components, density and temperature, field aligned currents, with $j_{\parallel i} = 0$ at $y_i = y = 0$ (as in *Lotko et al.* [1987]) and with maxima ranging from 0.8 to 2.4 μA/m^2 in the ionosphere. Also shown are profiles of B_z and $B_x|_{z=H}$ in the LLBL to illustrate diamagnetic effects and field-line bending, and the field-aligned potential drop (denoted by $\Phi_\parallel$ in the fig-

ure). The latter ranges from 0.8 kV to 2.8 kV. It is noted that, while the width of the region of tailward flow increases with x, the high-density/low temperature part of the LLBL actually becomes thinner as the flow speed increases. This is a direct consequence of mass conservation. Diffusive inflow across the magnetopause, which in principle is allowed in both the *Drakou et al.* and the *Wei et al.* formulations, would counteract this thinning. The potential drop across the LLBL observed in the simulations was found to reach values up to 10 kV.

Although large uncertainties and variations exist in parameter values deduced from observations, it appears that models of the type used by *Wei et al.* [1996] are reasonably consistent with what has been learned from a variety of direct and indirect observations of the LLBL and its ionospheric footprint.

4.3. Time-dependent flow in the LLBL

It was realized early on that, whether or not it occurs at the magnetopause itself, the Kelvin-Helmholtz (K-H) instability should be expected at the inner edge of the LLBL [*Sonnerup*, 1980; *Sckopke et al.*, 1981; *Lee et al.*, 1981] and that its effect would be to break up any initially smooth LLBL structure into a tailward travelling vortex train with an associated formation of LLBL plasma blobs. The topic of K-H instability in the LLBL and its consequences is beyond the scope of this review (for a sample of representative work, see [*Miura*, 1996], [*Keller and Lysak*, 1999], and [*Nykiri and Otto*, 2001]). Here we discuss briefly the work by *Wei and Lee* [1993]. They proposed that the ionospheric footprints of vortical structures in the LLBL would account for beads of UV bright spots observed in the post-noon sector, e.g., by the Viking spacecraft [*Lui et al.*, 1989]. They carried out a sophisticated time-dependent numerical simulation to verify this idea. Their basic configuration is similar to that used by *Wei et al.* [1996], e.g., the tailward flow velocity at the magnetopause increases with increasing distance from the subsolar point. They included a small diffusive mass flux across the magnetopause in order to populate the LLBL with magnetosheath plasma. The boundary layer approximation was not used because it does not permit studies of the vortical structures. The motion was treated as incompressible 2-D flow transverse to a uniform field. Viscous forces stemming from microscopic plasma instabilities were found to have a negligible effect for $\mu/\rho \leq 10^9$ m^2/s and the mapping from the equatorial region to the ionosphere was treated as isotropic ($dx/dx_i = dy/dy_i \simeq 35$) in order to simplify the analysis. Both Pedersen and Hall currents were

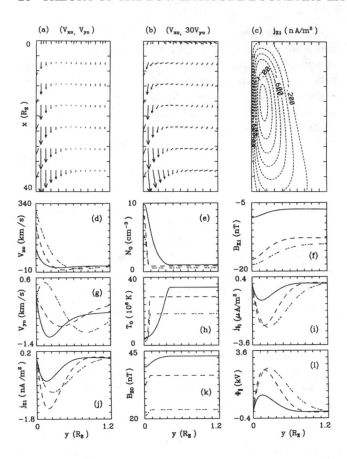

Figure 6. Results from LLBL model with constant viscosity, finite field-aligned potential drops, and nonconstant mapping factors. One-dimensional profiles of various quantities are shown at flow distances $-x = 4R_E$ (solid curves); $-x = 15R_E$ (dashed curves) and $-x = 30R_e$ (dot-dashed curves) [from *Wei et al.*, 1996].

included in the ionosphere, with conductivities that increase with increasing field-aligned potential difference, $\Delta\Phi_\| = \kappa(\Phi_e - \Phi_i)$, for upward currents ($j_{\|i} < 0$; $\Delta\phi_\| < 0$) and are constant ($\Sigma_{P0} = 0.5\Sigma_{H0} = 1$ mho) for downward currents ($j_{\|i} > 0$; $\Delta\Phi_\| > 0$), according to a model developed by *Kan and Lee* [1980]. Further, the field-aligned conductance, κ, was taken to be 10 times larger ($\kappa = 5 \times 10^{-8}$ mho/m^2) for downward currents than for upward currents ($\kappa = 5 \times 10^{-9}$ mho/m^2).

An example of the results for the post-noon LLBL from the *Wei and Lee* simulation is shown in Figure 7. The gradual growth of the average LLBL width with increasing distance from the subsolar point (which is at the top of the figure) is caused by the mass addition by diffusion across the magnetopause but this effect is counteracted to some extent by the gradual increase in tailward flow speed at the magnetopause. The density

blobs and the associated vortical motions are clearly seen as is the filamentation of $j_{\|i}$ and the generation of "hot spots" in the field-aligned power density $W_\| = j_{\|i}\Delta\Phi_\|$. On the pre-noon side, the behavior is similar, except that $W_\|$ is now negligibly small, i.e., no hot spots occur.

Two additional comments are needed. First, the description of the coupling of the LLBL used by *Wei and Lee* is quasistatic in the sense that it is applicable only when time variations are slow compared to the transit time of Alfvén waves from the equatorial region to the ionosphere. Also, their model does not incorporate the possible effects of field-line bending (see Section 4.1) on the turbulence.

The second comment is that, in the time-averaged description of the LLBL used in most of this review, the velocity fluctuations found by *Wei and Lee* would provide turbulent (Reynolds) stresses in the momentum equation that can in principle be expressed in terms of an eddy viscosity. The viscosity coefficient we have used in the steady-state models is an, admittedly oversimplified, way to represent these Reynolds stresses.

5. LLBL PARTLY ON OPEN FIELD LINES

The discussion up to this point has been concerned with the LLBL on closed field lines, i.e., field lines that have both feet in the ionosphere. But in the real magnetosphere, magnetic field reconnection is likely to be active somewhere on the magnetopause essentially all the time. And the reconnection process tends to consume any preexisting LLBL on closed field lines and convert it to a boundary layer on open field lines, i.e., field lines that have only one foot in the ionosphere, the other end being dragged tailward by the solar wind. Two situations will be mentioned, both of which may occur depending perhaps on the solar-wind parameters.

In the first scenario, the reconnection process at the magnetopause is patchy and highly intermittent. In that case, magnetospheric field lines that have been opened by reconnection may undergo a second reconnection to become closed again, as envisaged by *Kan* [1988] and *Nishida* [1989]. In this manner magnetosheath plasma may be transferred to closed LLBL field lines. The time-averaged effects of such multiple reconnections should be expressible in terms of suitable transport coefficients for mass diffusion and viscosity in the LLBL models presented in this review.

In the second scenario, shown in Figure 8, reconnection is quasisteady and occurs along an extended reconnection line, or X line for short. There is rapid

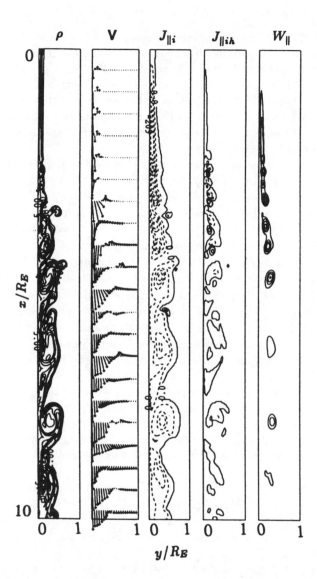

ρ V $J_{\parallel i}$ $J_{\parallel ih}$ $W_{\parallel}$

Figure 7. Contour plot of plasma density in the post-noon LLBL region, along with velocity vectors, contours of constant field-aligned current density, contours of the Hall portion of those currents and of the dissipation rate, $W_{\parallel} \equiv j_{\parallel i}\Delta\Phi_{\parallel}$ [from *Wei and Lee* 1993].

mentum equation, (21) or (41), for flow of LLBL plasma in the $-x$ direction because those stresses are exactly balanced by changes in the momentum of the magnetosheath plasma as it crosses the magnetopause. However, these stresses do modify the boundary conditions for the LLBL flow at its magnetopause edge.

We refer to the outer portion of the LLBL, which is threaded only by open field lines that do not pass

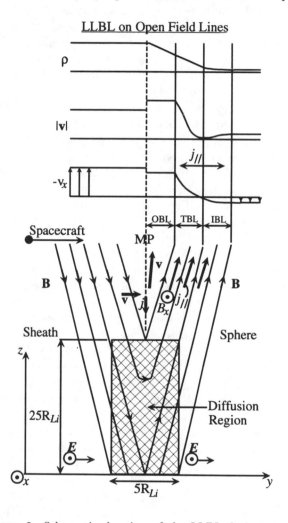

Figure 8. Schematic drawing of the LLBL during quasi-static reconnection. The wedge-shaped outer boundary layer (OBL) is where plasma jetting and various kinetic signatures of reconnection are mainly seen. The thin slab-shaped transition boundary layer (TBL) is threaded by open field lines that pass through the ion diffusion region; the similarly thin, inner boundary layer (IBL) is threaded by closed field lines that pass through the diffusion region. Signatures in plasma density, ρ, plasma speed, $|\mathbf{v}|$, and tailward velocity component, $-v_x$, as seen by a spacecraft traversing the magnetopause/LLBL region close to, but above, the ion diffusion region, are schematically shown.

tailward flow along the X line (in the $-x$ direction) in the magnetosheath and slow sunward flow in the magnetosphere. An LLBL on open field lines is formed. It contains mainly magnetosheath plasma that has crossed the magnetopause and been accelerated away from the X line by the Maxwell stresses in the magnetopause. This effect and associated kinetic signatures have been extensively studied and are discussed elsewhere in this monograph. The basic point to be made here is that the Maxwell stresses at the outer edge of the LLBL, i.e., at the magnetopause, do not enter into the mo-

through the diffusion region (a, perhaps inappropriately, named region in which the frozen field condition breaks down, first for ions and then for electrons, thus allowing reconnection to occur), as the outer boundary layer (OBL). As shown in Figure 8, the total velocity as well as its $-x$ component are approximately constant in this layer (see however *Levy et al.* [1964]), although they are not the same as in the magnetosheath; the density may decrease and the temperature increase gradually as one moves earthward in the layer as a result of mixing of magnetosheath and magnetospheric plasma populations. Viscous forces are small because of the small velocity gradients in the OBL so that the flow in the $-x$ direction is controlled by pressure gradients, dp/dx, and $\mathbf{j}_{00} \times \mathbf{B}$ forces, as discussed in Section 3.4. Figure 8 shows that, by definition, the width of the OBL is zero at the upper edge of the ion diffusion region; it is wedge shaped with wedge angle proportional to the reconnection rate. If the OBL is observed well away from the reconnection site, i.e., at a substantial z value in Figure 8, its width may be substantial. It may be much larger than the width of the remaining parts of the LLBL, discussed below.

Immediately earthward of the OBL, a thin transition layer (TBL) is located. This layer is threaded by open field lines that pass through the diffusion region. Its earthward edge is at the inner magnetic separatrix. The width of this layer is one half of the width of the ion diffusion region, i.e., it is at most a few ion gyroradii (R_{Li}) thick, and the width does not change much with the z coordinate. The velocity gradient $\partial v_x/\partial y$ is large in this region. Viscous stresses are therefore important and the field-aligned current densities $j_{\parallel}$ ($or j_{\parallel i}$) are expected to be large.

Adjoining the TBL on its earthward side is the inner boundary layer (IBL), which is threaded by closed field lines, which connect it to the inner part of the diffusion region. Velocity gradients and field-aligned currents are weaker in this layer. The density drops and the temperature rises to their respective magnetospheric levels in the IBL. The layer width is again very small — about 50% of the ion diffusion region width. It is essentially independent of the z coordinate.

The LLBL models developed in Sections 3 and 4 should be approximately applicable to the combined TBL and IBL region, which has a width equal to the diffusion region width, some $5R_{Li} \simeq 500$ km, say. Because of its small thickness, the velocity, density, and temperature changes across this region often appear very abrupt in time series of spacecraft data (see e.g., *Phan and Paschmann* [1996]). The outward drift of plasma

in the TBL and IBL, caused by the reconnection electric field, is counterbalanced by flow along open field lines in the TBL and by earthward diffusion. Note also that the OBL rather than the magnetosheath will provide the boundary condition on the velocity at the outer edge of the TBL. Included are z components that may cause the boundary layer height, H, to increase as one moves tailwards.

The physical processes in the diffusion region are in part rather different from those incorporated in our LLBL models but the total height of the upper half of the ion diffusion region is only some $25R_{Li} \simeq 2500$ km (for a dimensionless reconnection rate of 0.1 and $R_{Li} = 100$ km), which is a small fraction of the total effective height $H = 10R_E$ say, of the equatorial LLBL. In other words, the assumptions that were used in our LLBL models are applicable, at least approximately, in nearly the entire TBL + IBL region. The main difference between the present application and the application to a closed magnetosphere given in Table 1 is that the mapping factor dx/dx_i is smaller in an open magnetosphere: for $dx/dx_i = 60$ [*Wei et al.*, 1996] rather than 200, one finds $\delta_\nu \simeq 730$ km and $\lambda_e = 3870$ km. From equation (42) one then obtains $M_e = 5.3$. This M_e value indicates strong decoupling of the equatorial layer, principally the TBL, from the ionosphere and oscillatory behavior of the velocity and $j_{\parallel}$ profiles across the layer.

Our discussion of shear flow along an active X line has been mainly qualitative. Quantitative studies of such geometries are needed. A first attempt to numerically simulate the behavior of plasma and field was made by *Richard and Lotko* [1991]. Their model contains the combined effects of, on the one hand, reconnection, causing plasma flow toward the X line on both sides and accelerated flow along the magnetopause away from the X line (the z direction in Figure 8), and, on the other hand, strongly sheared viscous flow along the X line. The simulation shows that substantial B_x field components are generated in the boundary layer. These components are positive above the equatorial plane ($z > 0$) in Figure 8 and negative below that plane ($z < 0$). In other words, the sense of B_x is the same as that found in the self-consistent description of field-line bending in the LLBL, discussed in Sections 4.1 and 4.2. In Figure 8, the current system responsible for generating B_x in the OBL consists of an equatorward directed part in the magnetopause and a field-aligned Region 1 current heading toward the ionosphere in the TBL. In the Richard-Lotko simulation these currents are present but the transition through the magnetopause and boundary layer is smooth and gradual, i.e., the magnetopause does not appear as a thin layer, clearly separated from the TBL. Furthermore, there is no ionosphere in the simulation so that the Region 1 currents in the TBL are not providing a self-consistent connection to the ionosphere. Simulations that include such coupling would be desirable.

6. CONCLUSION

The models of the low-latitude boundary layer presented in this paper should serve as a guide in interpreting observations of this layer, both in the equatorial region and in the region of its ionospheric footprints. The model contains free parameters such as effective viscosity, mass diffusivity, field-aligned conductance, and mapping factors, which are difficult to predict from theory, but the values of which can be estimated, or at least placed within bounds, by comparison with observations. However, such studies are not simple to perform. The inward/outward motion of the magnetopause and the often intermittent nature of the LLBL flow make it difficult to extract quantitative information, such as average layer width, from observations of the equatorial LLBL. And it is even more difficult to extract information about mapping factors: global numerical simulations may be the only realistic approach. It is also difficult to develop a global view of the LLBL and its evolution in the flow direction from sparse spacecraft data. Short of Constellation-type missions, global simulations with enhanced grid resolution in the magnetopause/LLBL region again provide the best promise.

*Acknowledgments.*This work was funded by NASA's Sun-Earth Connection Theory Program (SECTP) under Grant NAG 5-8135 to Boston University and by NASA under Grants NASW-99014 to Mission Research Corporation (MRC), Nashua, NH and NAG5-7185 to Dartmouth College. The ISM simulation code was developed at MRC under contract with the Defense Threat Reduction Agency (DTRA), 8725 John Kingman Road, MS 6201, Ft. Belvoir, VA 22060-6201.

REFERENCES

Bauer, T. M., R. A. Treumann, and W. Baumjohann, Investigation of the outer and inner low-latitude boundary layers, *Annales Geophysicae, 19,* 1065, 2001.

Bird, R. B., W. E. Stewart, and E. N. Lightfoot, *Transport Phenomena, 2nd ed.,* J. Wiley & Sons, Inc., New York, 2001.

Chiu, Y. T., and J. M. Cornwall, Electrostatic model of a quiet auroral arc, *J. Geophys. Res., 85,* 543, 1980.

Coleman, P. J., Jr., A model of the geomagnetic cavity, *Radio Sci., 6,* 321, 1971.

Cowling, T. G., *Magnetohydrodynamics,* p. 11, Adam Hilger Ltd., Bristol, England, 1976.

Drakou, E., B. U. Ö. Sonnerup, and W. Lotko, Self-consistent steady-state model of the low-latitude boundary layer, *J. Geophys. Res., 99,* 2351, 1994.

Eastman, T.E., E. W. Hones, Jr., S. J. Bame, and J. R. Asbridge, The magnetospheric boundary layer: Site of plasma, momentum and energy transfer from the magnetosheath into the magnetosphere, *Geophys. Res. Lett., 3,* 685, 1976.

Fridman, M., and J. Lemaire, Relationship between auroral electron fluxes and field-aligned electric potential differences, *J. Geophys. Res., 85,* 664, 1980.

Galinsky, V. L., and B. U. Ö. Sonnerup, Dynamics of shear velocity layer with bent magnetic field lines, *Geophys. Res. Lett., 21,* 2247, 1994.

Iijima, T., and T. A. Potemra, The amplitude distribution of field-aligned currents at northern high latitudes observed by Triad, *J. Geophys. Res., 81,* 2165, 1976.

Kan, J. R., and L. C. Lee, Theory of imperfect magnetosphere-ionosphere coupling, *Geophys. Res. Lett., 7,* 633, 1980.

Kan, J. R., A theory of patchy and intermittent reconnection for magnetospheric flux transfer events, *J. Geophys. Res., 93,* 5613, 1988.

Keller, K. A., and R. L. Lysak, A two-dimensional simulation of the Kelvin-Helmholtz instability with magnetic shear, *J. Geophys. Res., 104,* 25,097, 1999.

Kelley, M. C., *The Earth's Ionosphere, Plasma Physics and Electrodynamics,* p. 39, Academic Press, Inc., San Diego, 1989.

Knight, S., Parallel electric fields, *Planet. Space Sci., 21,* 741, 1973.

Le, G., C.T. Russell, J. T. Gosling, and M. F. Thomsen, ISEE observations of low-latitude boundary layer for northward interplanetary magnetic field: Implications for cusp reconnection, *J. Geophys. Res., 101,* 27,239, 1996.

Lee, L. C., R. K. Albano, and J. R. Kan, Kelvin-Helmholtz instability in the magnetopause boundary layer region, *J. Geophys. Res., 86,* 54, 1981.

Levy, R. H., H. E. Petschek, and G. L. Siscoe, Aerodynamic aspects of the magnetospheric flow, *AIAA J., 2,* 2065, 1964.

Lotko, W., B. U. Ö. Sonnerup, and R. L. Lysak, Nonsteady boundary layer flow including ionospheric drag and parallel electric fields, *J. Geophys. Res., 92,* 835, 1987.

Lotko, W., and B. U. Ö. Sonnerup, The low-latitude boundary layer on closed field lines, in *Physics of the Magnetopause, Geophys. Monograph 90,* edited by P. Song, B. U. Ö. Sonnerup, and M. F. Thomsen, pp. 371-383, American Geophysical Union, Washington DC, 1995.

Lui, A. T. Y., D. Venkatesan, and J. S. Murphree, Auroral bright spots on the dayside oval, *J. Geophys. Res., 94,* 5515, 1989.

Lyons, L. R., Generation of large-scale regions of auroral currents, electric potentials, and precipitation by the divergence of the convection electric field, *J. Geophys. Res.*, *85*, 17, 1980.

Miura, A., Stabilization of the Kelvin-Helmholtz instability by the transverse magnetic field in the magnetosphere-ionosphere coupling system, *Geophys. Res. Lett.*, *23*, 761, 1996.

Nishida, A., Can random reconnection on the magnetopause produce the low-latitude boundary layer?, *Geophys. Res. Lett.*, *16*, 227, 1989.

Nykiri, K., and A. Otto, Plasma transport at the magnetospheric boundary due to reconnection in Kelvin-Helmholtz vortices, *Geophys. Res. Lett.*, *28*, 3565, 2001.

Phan, T.-D., B. U. Ö. Sonnerup, and W. Lotko, Self-consistent model of the low-latitude boundary layer, *J. Geophys. Res.*, *94*, 1281, 1989.

Phan, T.-D., and G. Paschmann, Low-latitude dayside magnetopause and boundary layer for high magnetic shear, 1. Structure and motion, *J. Geophys. Res.*, *101*, 7801, 1996.

Richard, R. L., and W. Lotko, Magnetic field draping at the low-latitude magnetopause, *J. Geophys. Res.*, *96*, 15,779, 1991.

Schlichting, H., *Boundary-Layer Theory*, McGraw-Hill Book Co., New York, 1968.

Sckopke, N., G. Paschmann, G. Haerendel, B. U. Ö. Sonnerup, S. J. Bame, T. G. Forbes, E. W. Hones, Jr., and C. T. Russell, Structure of the low-latitude boundary layer, *J. Geophys. Res.*, *86*, 2099, 1981.

Siscoe, G. L., W. Lotko, and B. U. Ö. Sonnerup, A high-latitude, low-latitude boundary layer model of the convection current system, *J. Geophys. Res.*, *96*, 3487, 1991.

Song, P., and C. T. Russell, Model of the formation of the low-latitude boundary layer for strongly northward interplanetary magnetic field, *J. Geophys. Res.*, *97*, 1411, 1992.

Song, P., D. L. DeZeeuw, T. I. Gombosi, and C. P. T. Groth, A numerical study of solar wind-magnetosphere interaction for northward interplanetary magnetic field, *J. Geophys. Res.*, *104*, 38,361, 1999.

Sonnerup, B. U. Ö., Theory of the low-latitude boundary layer, *J. Geophys. Res.*, *85*, 2017, 1980.

Sonnerup, B. U. Ö., K. D. Siebert, W. W. White, D. R. Weimer, N. C. Maynard, J. A. Schoendorf,, G. R. Wilson, G. L. Siscoe, and G. M. Erickson, Simulations of the magnetosphere for zero IMF: The groundstate, *J. Geophys. Res.*, in press, 2001.

Vasyliunas, V. M., The interrelationship of magnetospheric processes, in *Earth's Magnetospheric Processes*, edited by B. M. McCormac, pp. 29-38, D. Reidel, Hingham, Mass., 1972.

Wei, C.-Q., and L. C. Lee, Coupling of the magnetopause boundary layer to the polar ionosphere, *J. Geophys. Res.*, *98*, 5707, 1993.

Wei, C.-Q., B. U. Ö. Sonnerup, and W. Lotko, Model of the low latitude boundary layer with finite field-aligned potential drops and nonconstant mapping factors, *J. Geophys. Res.*, *101*, 21,463, 1996.

Woch, J., M. Yamauchi, R. Lundin, T. A. Potemra, and L. J. Zanetti, The low-latitude boundary layer at midlatitudes: Relation to large-scale Birkeland currents, *Geophys. Res. Lett.*, *20*, 2251, 1993.

Yang, Y. S., R. W. Spiro, and R. A. Wolf, Generation of region 1 current by magnetospheric pressure gradients, *J. Geophys. Res.*, *99*, 223, 1994.

B. Sonnerup, Thayer School of Engineering, Dartmouth College, 8000 Cummings Hall, Hanover, NH 03755, USA. (e-mail: sonnerup@dartmouth.edu)

K. Siebert, Mission Research Corporation, 589 West Hollis Street, Suite 201, Nashua, NH 03062, USA. (e-mail: ksiebert@mrcnh.com)

Modeling Magnetospheric Sources

Raymond J. Walker[1,2], Maha Ashour-Abdalla[1,3], Tatsuki Ogino[4], Vahe Peroomian[1], and Robert L. Richard[1]

We have used global magnetohydrodynamic simulations of the interaction between the solar wind and magnetosphere together with single particle trajectory calculations to investigate the sources of plasma entering the magnetosphere. In all of our calculations solar wind plasma primarily enters the magnetosphere when the field line on which it is convecting reconnects. When the interplanetary magnetic field has a northward component the reconnection is in the polar cusp region. In the simulations plasma in the low latitude boundary layer (LLBL) can be on either open or closed field lines. Open field lines occur when the high latitude reconnection occurs in only one cusp. In the MHD calculations the ionosphere does not contribute significantly to the LLBL for northward IMF. The particle trajectory calculations show that ions preferentially enter in the cusp region where they can be accelerated by non-adiabatic motion across the high latitude electric field. For southward IMF in the MHD simulations the plasma in the middle and inner magnetosphere comes from the inner (ionospheric) boundary of the simulation. Solar wind plasma on open field lines is confined to high latitudes and exits the tailward boundary of the simulation without reaching the plasma sheet. The LLBL is populated by both ionospheric and solar wind plasma. When the particle trajectories are included solar wind ions can enter the middle magnetosphere. We have used both the MHD simulations and the particle calculations to estimate source rates for the magnetosphere which are consistent with those inferred from observations.

1. INTRODUCTION

There are two main sources of magnetospheric plasma: the solar wind and the ionosphere. Classical theory of the

[1]Institute of Geophysics and Planetary Physics, University of California, Los Angeles, California

[2]Department of Earth and Space Sciences, University of California, Los Angeles, California

[3]Department of Physics and Astronomy, University of California, Los Angeles, California

[4]Solar Terrestrial Environment Laboratory, Nagoya University, Toyokawa, Aichi, Japan

Earth's Low-Latitude Boundary Layer
Geophysical Monograph 133
10.1029/133GM03

solar wind interaction with the magnetosphere predicts that the magnetopause should be an impenetrable boundary separating the shocked solar wind plasma of the magnetosheath from the hot tenuous plasma of the magnetosphere. However, three decades of observations have demonstrated clearly that magnetosheath plasma exists inside all regions of the magnetopause (see *Sibeck et al.* [1999] for a recent review). These magnetospheric boundary layers include the low-latitude boundary layer (LLBL), the polar cusp entry layer and the high latitude plasma mantle (PM).

Observations from low-altitude spacecraft indicate that shocked solar wind plasma from the magnetosheath enters the magnetosphere over a wide region under almost all solar wind conditions [e.g. *Newell and Meng,* 1992]. Observations in the magnetotail lobes indicate that plasma enters the magnetosphere along its entire length. Recently

Hultqvist et al. [1999] have combined the results from a large number of studies to identify and quantify the major source regions for magnetospheric plasma. They estimate that $\sim 10^{26}s^{-1}$ solar wind particles enter through the dayside magnetopause while 10^{28}-$10^{29}s^{-1}$ enters along the tail magnetopause. However most of the particles that enter through the tail magnetopause also exit via the tail and never reach the closed field line region of the magnetosphere where the required source strength is only a few times 10^{26}/s.

The ionosphere too contributes plasma to the magnetosphere. *Moore et al.* [1999] organized the published observations from polar orbiting spacecraft [*Yau et al.*, 1985; *Abe et al.*, 1996] to summarize the ionospheric outflow rates. Their compilation shows that the ionosphere can provide $\sim 10^{26}$/s of O^+ and H^+ to the magnetosphere.

A number of processes have been proposed to account for the transfer of solar wind plasma into the magnetosphere. These include magnetic reconnection, finite Larmor radius effects, diffusion, the Kelvin-Helmholtz instability, impulsive penetration and direct cusp entry. Over the past decade several modeling studies have investigated the sources of magnetospheric plasma. Several authors have used particle trajectory calculations in empirical electric and magnetic fields to investigate where both solar wind [*Ashour-Abdalla et al.*, 1993] and ionospheric sources [*Delcourt et al.*, 1992; 1994; *Peroomian et al.*, 1996] populate the magnetosphere while Richard and colleagues [*Richard et al.* 1994; *Walker et al.*, 1996] combined global magnetohydrodynamic (MHD) simulations with particle trajectory calculations to investigate the entry mechanisms for solar wind ions. More recently *Winglee* [1998; 2000] developed a multifluid MHD code with an ionospheric fluid. For northward interplanetary magnetic field he found that the boundary separating the region dominated by the ionospheric fluid from the region dominated by the solar wind (called the geopause by *Moore and Delcourt,* [1995]) was very close to the Earth while for southward IMF ionospheric plasma dominated the both inner and middle magnetosphere.

In this paper we use computer simulations to study the sources of plasma for the magnetosphere with emphasis on the sources of the LLBL. First we investigate the sources of plasma by using a global magnetohydrodynamic (MHD) simulation of the interaction of the solar wind with the magnetosphere. We use the MHD model to investigate both the ionospheric and solar wind sources and find that the single fluid simulation too has a strong IMF dependence on the solar wind and ionosphere sources with the solar wind source dominant for northward IMF and the ionosphere dominant for southward IMF. Then we add single particle trajectory calculations of solar wind ions to

the global MHD calculations in order to estimate the importance of phenomena such as finite Larmor radius effects not included in MHD. These calculations suggest that both sources contribute to the LLBL. For each approach we estimate the source rate. In section 2 we briefly describe the models and techniques used in this study. The simulation results for northward IMF are presented in section 3 while those for southward IMF are in section 4. Finally in section 5 we discuss these results in the context of magnetospheric observations and other models.

2. APPROACH

Since the details of the global magnetohydrodynamic simulation code used for this study have been described previously [*Ogino et al.*, 1992; 1994] we will describe it only briefly. We solve the resistive MHD equations and Maxwell's equations as an initial value problem on a rectangular grid. For the cases with purely northward and southward IMF we used a (402×102×102) points grid with a uniform mesh size of $0.4R_E$. For these calculations the simulation box extends from $30R_E \geq x \geq -130R_E$, $0 < y \leq -40R_E$, and $0 < z \leq 40R_E$. We also have included results from simulations with IMF $B_Y \neq 0$. For these cases we used a simulation box with 322×82×162 points and a mesh size of $0.5R_E$. In the Z direction the calculation domain is $-40R_E \leq z \leq 40R_E$. We solve the differential equations by using a modified version of the Leapfrog scheme that is a combination of the Leapfrog scheme and the two-step Lax-Wendroff scheme. The simulation parameters are fixed to solar wind values at the upstream edge of the simulation box with free boundary conditions at the sides and back. For all the simulations we used symmetry boundary conditions at $y = 0$ while for the quarter magnetosphere calculation we also used symmetry boundary conditions at $z = 0$. The inner (ionospheric) boundary was placed at $3.5R_E$. The solar wind velocity was 300km/s, the density was $5cm^{-3}$ and the temperature was $2 \times 10^{5\circ}K$.

The simulation for northward IMF was initialized by using an unmagnetized solar wind flow that lasted for up to three hours. Then a northward IMF was introduced into the flow. After four hours with northward IMF during which time a quasi-steady magnetosphere developed, the IMF was turned southward and the simulation was run for another four hours. For the simulations with $B_Y \neq 0$ the initialization period was followed by a five-hour interval with northward IMF. Then the IMF was rotated in 15° increments in the YZ plane with a dwell time at each direction of 15 minutes [*Walker et al.*, 1999]. While the dwell time was sufficient to enable us to determine the

Flow Streamlines and Pressure Contours
(B$_Z$>0,B$_Y$=0)

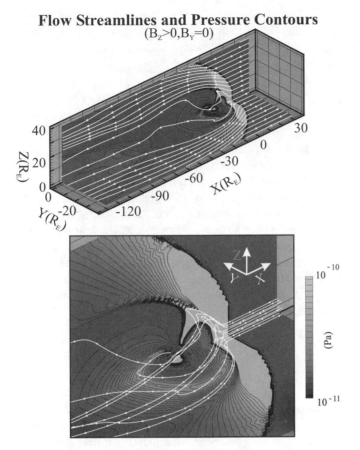

Figure 1. Flow streamlines and pressure contours for a simulation with IMF B$_Z$>0. The pressure contours were placed in the noon-midnight and equatorial planes. The flow streamlines were started in the solar wind. The bottom panel shows an enlargement of the region close to the Earth. Additional streamlines have been added near the Sun-Earth line.

entry mechanism, the overall magnetospheric configuration contains the effects of changing IMF orientations [*Walker et al.*, 1999]. For selected IMF directions the dwell time was increased to 2 hours to enable us to investigate the configuration of a quasi-steady magnetosphere. For all of the simulations the magnitude of the magnetic field was $|\vec{B}| = 5nT$.

We also examined the entry of solar wind ions into the magnetosphere by calculating the ion trajectories using a fixed magnetic and electric field model [*Ashour-Abdalla et al.*, 1993]. We use magnetic and electric fields from the MHD simulations. The electric field is given by $\vec{E} = -\vec{v} \times \vec{B} + \eta \vec{J}$ where $\vec{v}$ is the velocity, $\vec{B}$ the magnetic field, and $\vec{J}$ is the current density. An explicit resistivity (η) is included in the MHD simulations [*Ogino et al.*, 1994]. In each case a distribution of ions with the temperature used in the MHD simulations was launched in

the solar wind upstream of the bow shock. The ion trajectories were calculated by solving the equation of motion by using a fourth order Runge-Kutta method.

3. NORTHWARD IMF RESULTS

We started our studies by modeling plasma entry when the interplanetary magnetic field was northward. In Figure 1 we have plotted flow streamlines for a case with purely northward IMF. We show one quarter of the magnetosphere with pressure contours on the noon-midnight meridian and equatorial planes. The bottom panel contains a blow up of the region near the Earth. This simulation was run for 3 hours with no IMF and then for 4 hours with a northward IMF. This snapshot was taken 2 hours after the northward IMF entered the simulation box and after a quasi-steady magnetospheric configuration had developed [*Ogino et al.*, 1994; *Bargatze et al.*, 1999]. Most of the flow impinging on the magnetosphere is diverted around the obstacle (top). However, some plasma can enter the magnetosphere. In the bottom panel we have plotted additional streamlines from the region around the subsolar point. These streamlines enter the magnetosphere and move tailward on magnetospheric field lines. In Figure 2 we have reproduced two of these flow streamlines and calculated magnetic field lines along each streamline. The

Flow Streamlines and Magnetic Field Lines
(B$_Z$>0, B$_Y$=0)

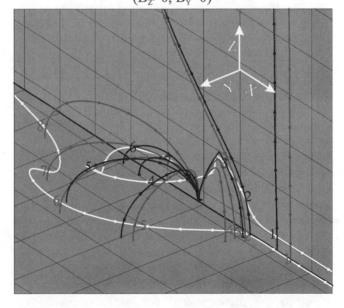

Figure 2. Magnetic field lines calculated along two flow streamlines from the simulation in Figure 1. The numbers indicate the locations along the streamlines discussed in the text.

Field Lines Along a Dusk LLBL Trajectory

(B$_Z$>0; B$_Y$>0)

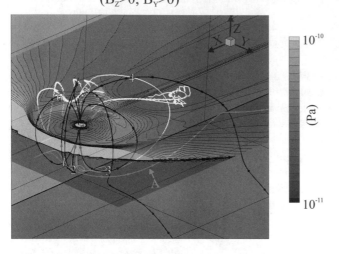

Figure 3. Field lines (black) along the trajectory of a solar wind ion (white). The trajectory was calculated in the electric and magnetic fields from a simulation with IMF B$_Z$>0 and B$_Y$>0. At this time B$_Z$=4.8nT and B$_Y$=1.3nT. Pressure contours have been plotted in the equatorial plane. They have been made semi-transparent. The gray line (A) is a flow streamline started from the same position as the ion trajectory. The numbers indicate locations along the trajectory described in the text.

black field lines correspond to a streamline along the noon-midnight meridian slightly northward of the equator. This streamline moves from solar wind to magnetospheric field lines at location 3 when the field line on which it is convecting reconnects in the polar cusp region forming a new closed magnetospheric field line. The streamline then moves around the flank magnetosphere into the near-Earth tail region. Note that the streamline remains on field lines that map to the LLBL while on the dayside. The second streamline starts along the noon-midnight meridian at the equator. The corresponding field lines are gray. Again the streamline enters the magnetosphere when the field line on which it is convecting reconnects with high latitude tail field lines. The streamline then moves along the dayside LLBL and into the tail.

The magnetospheric configuration changes significantly when the IMF has a B$_Y$ component. In the MHD model for purely northward IMF, the IMF field lines reconnect simultaneously in the northern and southern hemisphere forming closed field lines. When there is a B$_Y$ component the reconnection can occur in only one hemisphere. In Figure 3 we have plotted pressure contours in the equatorial plane along with selected magnetic field lines from a snapshot from a simulation when the IMF had northward and dawnward components. At this time the IMF had been northward for 45 minutes and duskward for

90 minutes. The pressure contours have been made transparent so that field lines (in black) south of the equator can be seen. Note the open field lines shown here along the dusk flank. A flow streamline starting in the southern hemisphere just duskward of noon has been drawn in gray (A). As in the purely northward IMF case, this streamline enters the magnetosphere at approximately position 2 when the field line on which it is convecting reconnects at high latitude. The streamline enters the magnetosphere on open field lines and remains on open field lines until it exits the simulation box.

We also launched a distribution of solar wind test particles in the magnetic and electric fields from this MHD simulation. The distribution function of the test particles was selected to have the same plasma moments as the MHD simulation. The test particles were all protons and were launched in the plane at $x = 20R_E$. The white trace in Figure 3 gives the trajectory of one of these ions. This proton has energy of about 190eV in the solar wind frame. The ion entered the magnetosphere at approximately position 2. Again the entry mechanism is high latitude reconnection. Using similar calculations *Richard et al.,* [1994] determined that high latitude reconnection is the main way in which solar wind ions enter the magnetosphere when the IMF is northward. After entering the magnetosphere the ion bounced several times at high southerly latitudes before moving onto open field lines and then into the polar cusp region. The ion emerged from the cusp onto closed dayside field lines. In Figure 4 this trajectory has been plotted from two perspectives. The view in the top panel is from the afternoon while the view in the bottom is from the morning. Field lines along the trajectory have been shaded to indicate the magnetic field magnitude. Note that soon after the ion crossed the magnetopause it bounced about the off equatorial minimum $|\vec{B}|$ that is characteristic of the magnetosphere just earthward of the dayside magnetopause. This type of oscillation about the high-latitude minimum $|\vec{B}|$ has been reported by *Delcourt et al.* [1992] based on particle calculations in empirical $\vec{E}$ and $\vec{B}$ field models (see their Figure 4). However, by the time the ion in Figure 4 reached the morning side (bottom) it was on a trapped orbit bouncing about the equator. In between the ion moved straight across the dayside cusp near the region where the high latitude reconnection is occurring.

The effects of the motion across the cusp can be seen in Figure 5. Here the trajectory has been projected onto three orthogonal planes. Each trajectory is shaded to show the energy of the particle. While the ion is on the dusk side it has about the same energy as it had in the magnetosheath. However, when it moved across the polar cusp it quickly gained ~20kV. *Richard et al.,* [1994] argued that ions

Field Lines Along a Dusk LLBL Ion Trajectory
($B_Z > 0$; $B_Y > 0$)

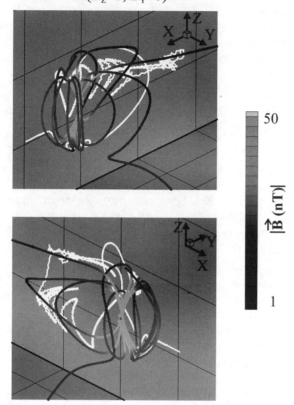

Figure 4. Two views of the ion trajectory in Figure 3. The view in the top panel is from the dusk side while the view in the bottom panel is from the dawn side. Magnetic field lines have been calculated along the trajectory. The gray scale on the field lines gives the magnitude of the magnetic field.

move non-adiabatically in the polar cusp region when high latitude reconnection causes the radius of curvature of field lines there to become comparable with the ion gyro-radius. This non-adiabatic motion carries the ions across the high latitude electric field where they are accelerated.

The ion in Figure 5 is unusual in two ways. First it gained more energy in the cusp region than typical ions. Second it entered the cusp region after moving in the dusk side magnetosphere. However, our trajectory calculations indicate that non-adiabatic motion across the polar cusp region is the main way in which the most energetic ions are accelerated. Most of the accelerated ions enter the cusp directly from the magnetosheath when the field line on which they are moving reconnects at high latitudes.

Ions that do not move through the high latitude reconnection region gain much less energy. We have plotted the trajectory of an ion that did not enter the high latitude reconnection region in Figure 6. The trajectory has

gray shading to show the energy of the particle. This ion entered the magnetosphere very close to noon when the field line on which it was moving reconnected with the southern lobe field (not shown). It quickly moved to closed field lines and bounced and drifted near the dayside magnetopause to the night side. This ion gained less than one keV as it moved across the dayside magnetosphere.

In Table 1 we have estimated the ion entry rates for the northward IMF simulations. We estimated the entry rate in the MHD simulation by determining the streamlines that cross the magnetopause onto magnetospheric field lines and calculating $\int \rho \, \vec{v} \cdot d\vec{s}$ where ρ is the mass density, $\vec{v}$ is the fluid velocity and $d\vec{s}$ is an area element on the magnetopause. The integral was taken over the magnetopause area threaded by penetrating streamlines. Since the variable in the MHD simulations is the mass density, we assumed the solar wind to consist solely of

Ion Energy Along Trajectory

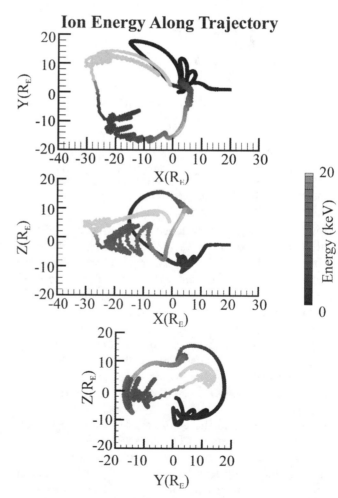

Figure 5. Projections of the ion trajectory in Figure 3 onto three planes (Z=0, Y=0, X=0). The gray scale gives the particle energy.

Ion Energy Along Trajectory

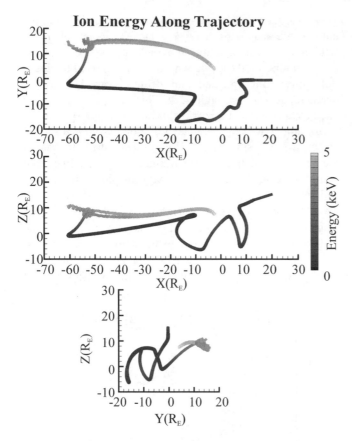

Figure 6. Projections of an ion trajectory onto three planes (Z=0, Y=0, X=0). The gray shading gives the particle energy. The trajectory was calculated in the electric and magnetic fields from the same simulation as the particle in Figure 3.

protons in order to obtain the entry rate. When the IMF is purely northward 4×10^{26} /s enter on closed field lines. When the IMF has a B_Y component we have listed two numbers. The first is the entry rate onto closed field lines. It remains about the same as in the case with $B_Y = 0$. The second number is the entry rate on both open and closed field lines. It is an order of magnitude larger because of ions that are inside the magnetopause but on open field lines. We used a simulation with $B_Y = -3.5nT$ and $B_Y = 3.5nT$ that was run until a quasi-steady magnetosphere was formed (2 hours) to estimate the entry rates for the $B_Y \neq 0$ case.

We also estimated the entry rate by using the particle trajectory calculations. The entry rate is given by $r = C(A \times v_{SW} \times n_{SW})/N$ where C is the number of ions entering the magnetosphere, N is the total number launched, A is the area in the solar wind over which the ions were launched, v_{SW} is the solar wind velocity and n_{SW} is the solar wind number density. In general the source rate estimates are larger from the trajectory calculations. In Figure 7 we have plotted the location on the magnetopause

where the ions entered the magnetosphere. The entry points are bunched near the polar cusp. In MHD the comparable distribution has more uniform entry. We get a higher estimate of the source rate from the particle calculations because of the enhanced entry in the cusp that is caused by the non-adiabatic motion [*Richard et al.,* 1994].

4. SOUTHWARD IMF RESULTS

We have plotted flow streamlines and pressure contours for a case with purely southward IMF in Figure 8. This snapshot was taken 60 minutes after the IMF was turned southward from northward. By this time subsolar point reconnection had been occurring for approximately 45 minutes [*Walker et al.,* 1993; *Bargatze et al.,* 1999]. The magnetosphere had not reached a steady state. A few minutes (t=70 minutes) after this snapshot reconnection began in the near-Earth ($x \approx -15R_E$) magnetotail. About 10 minutes later reconnection began on lobe field lines. Since the magnetospheric configuration was not steady we calculated flow streamlines through a series of time steps starting before and extending after this time. Since the results were very similar to those in Figure 8 we have not included them in this paper.

In the simulation flow streamlines from the solar wind do not reach the inner parts of the magnetosphere. The streamlines that reach the middle and inner parts of the magnetosphere come from the inner boundary of the

	TABLE 1. Ion Entry Rates $B_Z > 0$	
	MHD	Ion Trajectory
$B_Y = 0$	4×10^{26}/s	1×10^{27}/s
$B_Y \neq 0$	5×10^{26}/s , 5×10^{27}/s	2×10^{27}/s, 5×10^{27}/s

Ion Entry Locations

(B$_z$>0)

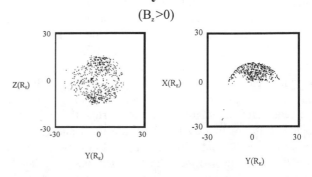

Figure 7. The distribution of ion entry locations on the dayside magnetopause. The left panel shows the entry locations projected onto the YZ plane while the right panel shows the locations projected onto the XY plane. The MHD simulation had IMF B$_Z$>0. [After Richard et al., 1994].

WALKER ET AL. 39

Flow Streamlines and Pressure Contours
($B_Z<0$, $B_Y=0$)

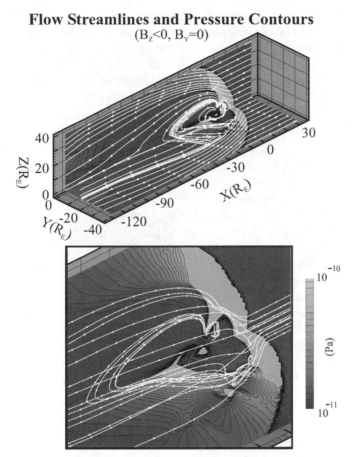

Figure 8. Flow streamlines and pressure contours for a simulation with IMF $B_Z<0$. The pressure contours were placed in the noon-midnight and equatorial planes. The flow streamlines were started in the solar wind and in the middle and inner magnetosphere. The bottom panel shows an enlargement of the region close to the Earth. Additional streamlines have been added near the Sun-Earth line.

simulation. Streamlines in the LLBL come from both the solar wind and the inner boundary. Those nearest the Earth come from the inner boundary while streamlines nearer the magnetopause come from both sources. Note that the earthward flowing streamlines from the inner boundary turn around in the dayside magnetosphere and join with streamlines from the solar wind in flowing tailward. This can be seen more clearly in Figure 9. Here we have plotted only three streamlines along with magnetic field lines along the streamlines. The two sources of the LLBL have dashed and solid black field lines. Plasma on the dashed field lines enters the magnetosphere when the field line reconnects at local times away from noon (white 2). This plasma moves on open field lines through the LLBL into the tail (white 3-5). In this example plasma from the inner boundary (solid black field lines) has taken a less direct

path to the LLBL. The streamline emerges from the inner boundary on the dayside (black 1) and moves toward the dayside magnetopause. There it is picked up by reconnection (black 2) and moves on open field lines into the tail (black 3-5). In the tail it is picked up by earthward convection (black 6) and moves back toward the dayside magnetosphere (black 6-8). At the dayside magnetopause (between black 8 and 9) the streamline again encounters reconnection. Finally it moves tailward on open field lines (black 10). In this example we followed a flow streamline from the inner boundary on the dayside. However plasma also emerges from the inner boundary on the nightside (Figure 8). These streamlines come from auroral latitudes. When the nightside flow streamlines reach the equator they too flow earthward into the dayside magnetosphere. Most of the outflow at the inner boundary is field aligned.

The gray field lines in Figure 9 have been calculated along a streamline that originates in the solar wind along the Earth-Sun line. It too enters the magnetosphere because of dayside reconnection (gray 2) but then moves on open field lines the length of the simulation box (gray 3-5).

We also launched a distribution of ions from the solar wind into the electric and magnetic fields from this simulation [*Walker et al., 1996*]. We have plotted an example showing the trajectory of one of those particles in

Flow Streamlines and Magnetic Field Lines
($B_Z<0$, $B_Y=0$)

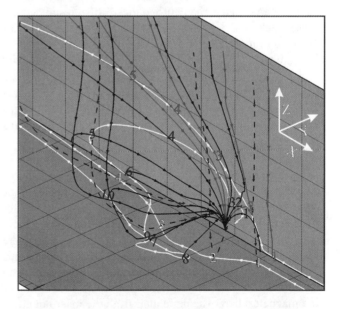

Figure 9. Magnetic field lines calculated along three flow streamlines from the simulation in Figure 8. The numbers indicate the locations along the streamlines discussed in the text.

Kelvin-Helmholtz Instability and Magnetic Reconnection: Mass Transport at the LLBL

A. Otto and K. Nykyri

Geophysical Institute, University of Alaska, Fairbanks, Alaska.

During periods of southward IMF magnetic reconnection appears to be the dominant process for the transport of magnetic flux and magnetosheath plasma into the magnetosphere. However, for northward IMF the transport of mass into the magnetosphere is not well understood. In particular for strongly parallel magnetic fields on the two sides of the magnetopause, magnetic reconnection does not operate or is inefficient for the mass transport. For largely parallel fields the Kelvin-Helmholtz (KH) mode is unstable for k vectors largely perpendicular to the magnetic field. Although the KH mode is an ideal instability, the deformation of the boundary magnetic field through the KH can generate strong current layers and force magnetic reconnection within the KH vortex motion. We will discuss this process, onset conditions, and implications for the structure of the Low Latitude Boundary Layer (LLBL). The transport of mass appears consistent with the diffusion required at the LLBL and with the typical correlation time for the magnetospheric plasma density in response to changes in the solar wind.

INTRODUCTION

Magnetic reconnection and Kelvin Helmholtz (KH) modes are of fundamental importance for many space plasma systems. Magnetic reconnection is discussed in active galactic nuclei, galactic jets, stellar atmospheres, and planetary magnetospheres. For the Earth's magnetosphere it is believed that reconnection plays an important role during magnetospheric substorms and provides a major mechanism for the transport of plasma, energy, and magnetic flux at the magnetospheric boundary preferably for southward interplanetary magnetic field (IMF).

Applications of the KH mode range from the flow around obstacles like cars or aircraft and atmospheric dynamics to space physics processes like flow of intergalactic matter and flow around planetary magnetospheres. The KH mode can provide the transport of momentum and energy from the so-

lar wind into the Earth's magnetosphere and many in situ observations indicate that the KH mode operates at the Earth's magnetospheric boundary. In a collisionless plasma and on scales much larger than the ion inertia scale the KH mode is an ideal instability, i.e., it cannot transport mass across magnetic field lines or alter the magnetic topology.

KH modes and magnetic reconnection are both driven by spatial inhomogeneities typical for the interaction of plasmas of different origin. The KH mode is driven by a gradient or shear in the plasma velocity as illustrated in Figure 1 which causes wave motion and velocity vortices (in the nonlinear state) at the plasma boundary. However, the presence of a magnetic field component aligned with the **k** vector of the instability stabilizes the mode because it requires additional energy to twist the magnetic field in the boundary or vortex motion. The onset condition for the KH mode is

$$\frac{m_0 n_1 n_2}{n_1 + n_2} [\mathbf{k} \cdot \Delta \mathbf{V}]^2 > \frac{1}{\mu_0} \left[(\mathbf{k} \cdot \mathbf{B}_1)^2 + (\mathbf{k} \cdot \mathbf{B}_2)^2 \right]$$

where m_0 is the ion mass, n is number density, $\Delta \mathbf{V} = (\mathbf{V}_1 - \mathbf{V}_2)$ is the velocity shear, **B** is magnetic field strength, and the indices denote plasma properties on the two

Earth's Low-Latitude Boundary Layer
Geophysical Monograph 133
Copyright 2003 by the American Geophysical Union
10.1029/133GM05

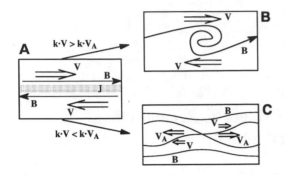

Figure 1. Sketch of magnetic reconnection and KH modes as a result of different initial conditions

sides of the boundary. In a plasma with constant density and constant $|\mathbf{k} \cdot \mathbf{B}|$ this relation becomes

$$(\mathbf{k} \cdot \mathbf{V}_0)^2 > (\mathbf{k} \cdot \mathbf{v}_A)^2$$

with $\mathbf{V}_0 = \frac{1}{2}(\mathbf{V}_1 - \mathbf{V}_2)$. Thus the velocity shear along $\mathbf{k}$ must be larger than the Alfvén speed along $\mathbf{k}$ for instability. Note that $\mathbf{B}$ can reverse along $\mathbf{k}$.

Magnetic reconnection requires anti-parallel field components across a current sheet (Figure 1). However, reconnection is entirely switched off if the shear velocity is larger then the Alfvén speed for the following reason. The information that reconnection operates propagates with Alfven speed along the magnetic field. If the plasma flow is faster than the local Alfvén speed this information cannot propagate away from the reconnection side. Properties of reconnection and KH modes alter in asymmetric plasma conditions but the qualitative physics remains the same.

Thus, in two dimensions either magnetic reconnection or the KH mode can operate for an initial condition with sheared plasma flow and anti-parallel field components. The KH mode amplifies the magnetic field in the KH vortex motion if an initial field component along the $\mathbf{k}$ vector is present. This amplification is significant in the nonlinear state of the instability. It is easy to see that a magnetic field which is wrapped up by the vortex motion must generate multiple current layers in the vortex as indicated in Figure 1. Velocity shear across the individual current sheets in the KH vortex is much less than the velocity shear of the initial configuration. Therefore the individual current layers can be unstable to the tearing mode and magnetic reconnection.

Observational evidence for the KH mode at the low latitude boundary layer (LLBL) has been reported by various authors [*Belmont and Chanteur*, 1989; *Fujimoto and Terasawa*, 1994; *Lee and Olson*, 1980; *Miura and Pritchett*, 1982; *Miura*, 1984; *Wu*, 1986]. A rather convincing case for the presence of the KH instability has been presented by *Fairfield et al.* [2000]and *Otto and Fairfield* [2000] to which

we will refer as F1 and O1 in the following. On March 24, 1995, the Geotail spacecraft observed large fluctuations of the magnetic field and plasma properties at the dusk side magnetospheric flank. At the time the spacecraft moved from the magnetosheath into the LLBL about 15 R_E tailward of the dusk meridian. A remarkable property of the observations reported by F1 is a strongly fluctuating magnetic field with brief periods of negative B_z components, although the interplanetary magnetic field (IMF) remained strongly northward and largely parallel to the plasma sheet magnetic field during the entire event. The field fluctuations and the plasma properties show a clear quasi-periodic behavior with a period of 2–3 min. The quasi-periodic signatures were observed through the entire transition of the LLBL lasting for about 5 hours during which the IMF remained fairly constant with a strong northward field component.

Quiet geomagnetic periods with presumably northward IMF have long been associated with a cold and dense plasma sheet [e.g., *Fairfield et al.*, 1981; *Lennartsson*, 1992]. *Borovsky et al.* [1998] demonstrated that the plasma sheet temperature and density are correlated to solar wind properties on a time scale of 1 to 2 hours. Recently *Terasawa et al.* [1997] and *Fujimoto et al.* [1998, 2000] reported a strong correlation between a high density cold plasma sheet and a strongly northward IMF orientation during the hours prior to the plasma sheet observations. *Lennartsson* [1992] showed that the plasma sheet composition is strongly dominated by the solar wind. It has been suggested that magnetosheath particles penetrate through the magnetospheric flank boundaries. However, except for cusp magnetic reconnection [*Song and Russell*, 1992] no specific mechanisms are suggested.

The following section presents results of two-dimensional MHD simulations studying signatures of KH waves. The basic configuration considers a strongly northward IMF and is illustrated in Figure 2c. The next section studies magnetic reconnection in KH vortices and the associated plasma transport. Section 4 addresses KH modes and magnetic reconnection in three dimensions and the last section presents a discussion of the results with application to the magnetospheric flanks.

SIGNATURES AND PROPERTIES OF KH MODES IN 2D

The following results apply to the presence of KH modes at the dusk side flank for strongly northward IMF. The KH mode is a well understood instability in two dimensions [*Belmont and Chanteur*, 1989; *Chen et al.*, 1997; *Fujimoto and Terasawa*, 1994, 1995; *Keller and Lysak*, 1999; *Lee and Olson*, 1980; *Miura and Pritchett*, 1982; *Miura*, 1984, 1987, 1992, 1999; *Thomas and Winske*, 1993; *Wilber and Winglee*, 1995; *Birk et al.*, 1999]. However, in most studies the magnetic field is assumed perpendicular to the

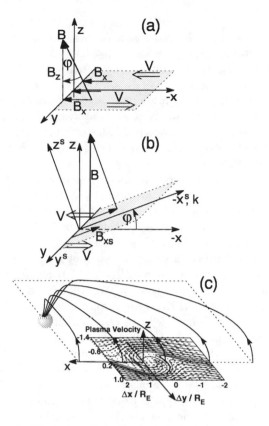

Figure 2. Illustration of the magnetic field geometry (a) and (b) and the KH mode propagating along the dusk side magnetospheric flank.

k vector of the KH mode. For real plasmas this is a singular situation. Usually the magnetic fields on the two sides of a plasma boundary are not exactly aligned. Even in cases where they are closely aligned (almost parallel for the March 24, 1995 events) the growth rate is almost unchanged for a small component of the magnetic field along the **k** vector of the instability. The basic configuration as it applies to the flank magnetospheric boundaries is illustrated in Figure 2. Here x points sunward along the boundary, y is the direction normal to the boundary, and z completes the coordinate system.

Figure 2a depicts a geometry with a small B_x component and the **k** vector is aligned with the magnetosheath flow direction. Figure 2b shows a case where the magnetic field is exactly in the z direction but the **k** vector is slightly tilted out of the x, y plane. In terms of the growth rate and stability these cases are rather similar. The general case can be expected to be a mixture of the two configurations.

Numerical Method

Most of the simulations presented here use a two-dimensional MHD model [*Otto*, 1990] in x and y with the geometry of Figure 2a. The initial configuration in the boundary coordinates uses a magnetic field of $B_{x0}(y) = b_0(y) \sin \varphi$, $B_{y0}(y) = 0$, and $B_{z0}(y) = b_0(y) \cos \varphi$ such that $90° - \varphi$ is the angle between the unperturbed magnetic field direction and the **k** vector of the KH mode. The velocity is $v_{x0} = v_0(y)$, $v_{y0} = v_{z0} = 0$. Initial density, pressure, velocity, and magnetic field magnitudes are

$$\rho_0 = \frac{1}{2}(\rho_{sh} + \rho_{sp}) + \frac{1}{2}(\rho_{sh} - \rho_{sp})\tanh\frac{y}{L_0}$$

$$p_0 = \frac{1}{2}(p_{sh} + p_{sp}) + \frac{1}{2}(p_{sh} - p_{sp})\tanh\frac{y}{L_0}$$

$$v_0 = -\frac{1}{2}V_{sh}\left(1 + \tanh\frac{y}{L_0}\right)$$

$$b_0 = \frac{1}{2}(b_{sh} + b_{sp}) + \frac{1}{2}(b_{sh} - b_{sp})\tanh\frac{y}{L_0}$$

where the indices sp and sh indicate the magnetosphere and magnetosheath. The parameters are chosen to model the Geotail 1995 events [F1, O1] and are listed in Table 1.

The simulations employ a unit length of $L_0 = 600$ km, typical magnetic field $B_0 = 16$ nT, number density $n = 11$ cm^{-3}, velocity (Alfvén speed) $V_A = 105$ km/s, and time scale $\tau_A = 5.7$ seconds. Periodic boundary conditions are used along the x direction and open boundary conditions are at y_{min} and y_{max}. An incompressible velocity perturbation with an amplitude of $0.01\,V_A$ is applied initially. All simulations are carried out in a frame moving with half the sheath velocity along the magnetospheric boundary.

We have chosen various values of φ but most results use $\varphi = 25°$. Figure 3 shows a plot of the dispersion relation

$$q = [\alpha_1\alpha_2[(\mathbf{V_1}-\mathbf{V_2})\cdot\mathbf{k}]^2 - \alpha_1(\mathbf{V_{A1}}\cdot\mathbf{k})^2 - \alpha_2(\mathbf{V_{A2}}\cdot\mathbf{k})^2]^{1/2}$$

as a function of φ and of V_{sh}. Black squares show simulation results compared to the analytic equation for an arbitrarily thin boundary. The figure demonstrates that the growth is almost unaltered for a large range of φ and is only moderately sensitive to the sheath velocity.

Magnetic and Plasma Signatures

A detailed account of the overall dynamics and the signatures obtained from KH simulations is found in O1. Here we will only consider the most important observational aspects. The example in Figure 4 shows magnetic field and plasma

Table 1. Initial Plasma Parameters

Location	B, nT	V, km s^{-1}	ρ, cm^{-3}	β
M'sphere	16	0	2.8	2.8
M'sheath	24	315	19.2	0.69

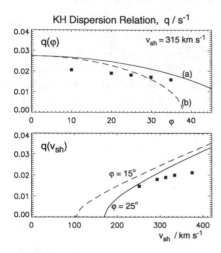

Figure 3. Dispersion relation as a function of φ (top) and of V_{sh} (bottom). Labels (a) and (b) refer to the corresponding configurations in Figure 1.

velocity of a KH simulation, projected into the x, y plane at time=358 seconds. B_z and plasma density are shown as gray scale plots. The vertical line indicates the slice along which plasma parameters will be studied.

In the plots of Figure 4 the magnetosheath is on the right and the vortex is moving in the negative x direction. A satellite at rest with the magnetosphere would move through the KH wave from the top in a time of approximately 2 minutes. The wave is changing sufficiently slow such that a slice through the wave for instance along the vertical line in Figure 4 is a reasonable representation of the typical signatures. The data which would be recorded is presented in Figure 5.

Figure 5 is a cut at $y = -3$ (-1800 km). Different shadings in Figure 5 indicate regions with different characteristic properties. The lighter shading indicates a region with high sheath-like density, the darker shading indicates a region of high temperature and low density, and the vertical lines indicate an additional boundary. We labeled distinct intervals using lettering from a to e.

Region a represents the core of the KH vortex and is characterized by strong fluctuations in all plasma and field properties. The region can show high-density spikes somewhat resembling sheath-like properties (higher B_z, etc.). Often visible in this region are reversals of V_y which can be accompanied by changes in V_x.

Region b shows fairly steady plasma and field properties, with magnetospheric-like high temperatures and relatively low densities. Characteristic of this region is also a fairly steady decrease of the V_y component.

Region c represents the outbound transition just prior to the high-density intervals. Typical of region c are the decrease in temperature and increase in density, a high total magnetic field magnitude, a pronounced minimum in B_z,

and extrema in the B_x and B_y components. The local maxima and minima in B_x and B_y are often accompanied by a minimum of B_z prior to the sheath-like intervals with larger and steady B_z. The outbound pass is often easy to identify in the Geotail data [F1].

Region d shows the already mentioned high number density and low temperature, large B_z, and steady plasma and field properties, i.e., magnetosheath-like properties. The length of high-density intervals decreases with distance from the magnetosheath.

Finally, region e marks the inbound pass into a region of generally lower number density and intermediate or high temperature. It shows a less pronounced minimum in B_z and frequently an extremum in B_y (sometimes in B_x depending on the normal direction of this boundary).

Figure 4 illustrates a strong depletion of B_z outside the vortex close to the original boundary (the dark region in the gray scale plot of B_z in Figure 4). We will address this region as the spine of the KH wave. A rotation back into GSM coordinates often shows negative B_z values in the center of this spine. Typical are also large B_x and B_y components in the spine region and it extends partially into the vortex. Such a region can also been identified in prior work [*Miura*, 1984, 1987; *Wu*, 1986]. An outbound pass has to cross this spine and would show typical extrema in B_x and B_y (with opposite polarity on the dusk side flank). The KH vortex motion leads to a twisting of the magnetic field, which accounts for the strong variations of the field in the vortex (region a). Figure 4 shows high-density plasma in the vortex explaining the spikes of high number density in region a.

The large magnetic field magnitude along the spine and the depletion of the B_z component in the same region is explained as follows. During the early evolution the KH vortex region develops a strong depletion of the static total pressure

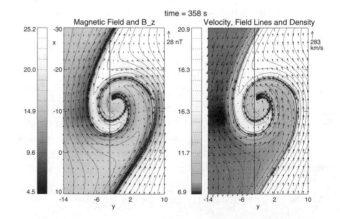

Figure 4. Snapshot of magnetic field lines, field vectors, and B_z (grayscale) on the left and plasma velocity (arrows), field lines, and plasma density (grayscale) on the right at 358 seconds into the evolution of the KH mode.

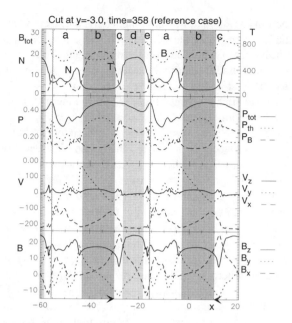

Figure 5. Data along the slices at $y = -3$ of the configurations shown in Figure 4

due to centrifugal forces of the vortex motion. Plasma is swept into the vortex along the spine region as illustrated by the large number of plasma elements (asterisks) in the vortex in Figure 4. The resulting plasma depletion along the spine region (outside the vortex) leads to the increase of the magnetic flux aligned with this boundary (B_x and B_y). The B_z component along the spine decreases at the same time because the frozen-in condition implies that B_z flux is swept into the vortex together with the plasma elements.

This evolution eventually leads to a stabilization of the vortex motion because the increasing field along the spine requires a larger energy in order to twist the magnetic field. It is expected and indicated by our results that the stabilization occurs when the magnetic field energy density (based on the B_x and B_y components) approaches the energy density of the shear velocity in the boundary region. This leads locally to an approximate equipartition between kinetic energy of the bulk flow and the magnetic energy, which may be relevant for astrophysical systems [*Birk et al.*, 1999]. The similarity of the simulation results and the actual Geotail observations is noteworthy and leaves little doubt that the quasi-periodic Geotail events on March 24, 1995 [F1] are caused by nonlinear KH waves moving across the spacecraft.

RECONNECTION IN NONLINEAR KH WAVES

Two-Dimensional Results

As demonstrated magnetic flux can be wrapped up in the KH vortex motion if the initial magnetic field has a small component along the **k** vector of the wave. During this pro-

cess strong current sheets evolve within the vortex (dotted and dashed lines in Figure 6). Magnetic reconnection can now occur in these current sheets and detach plasma filaments from one side of the original boundary as illustrated in the second and third sketch in Figure 6. The process has been reported in simulation studies [*Keller and Lysak*, 1999; *Otto and Fairfield*, 2000] and is analyzed quantitatively by *Nykyri and Otto* [2001].

This mechanism can transport plasma across the initial boundary even though the original magnetic field is parallel on the two sides of the initial boundary. Thus it may be of particular importance for the plasma transport from the magnetosheath into the magnetosphere during periods of strongly northward IMF. Properties of this process are studied with 2D MHD simulations in the configuration described in section 2. The presented results use values for φ ranging from 3° to 35°. Asymmetry across the flow boundary must be present in the initial configuration to facilitate a net transport of plasma. This asymmetry is provided by the strong density gradient between the magnetosheath and the magnetosphere. Asymmetry is also introduced by a different magnetic field strength (Table 1). To study the effect of the magnetic field asymmetry we reversed the field asymmetry in one set of simulation cases for the same values of φ.

Figure 7 illustrates the evolution of a KH wave for $\varphi = 10°$ in terms of plasma velocity, magnetic field, and density. The white line marks the initial plasma boundary and asterisks indicate the location of plasma elements which were initially uniformly distributed along the boundary. The figure illustrates the formation of a flux rope (magnetic island) shortly before $t = 74.5 \tau_A$ and further flux ropes are formed by the time $t = 89.4 \tau_A$ (note that the boundary condition along x is periodic). These newly formed flux ropes consist of plasma which was originally on the 'magnetosheath' side of the flow boundary and are now embedded entirely in 'magnetospheric' plasma as sketched in Figure 6.

To quantify the mass transport associated with the reconnection process the plasma density is integrated over the area of the islands and recorded as a function of time. The islands are identified by magnetic field lines marked by the plasma

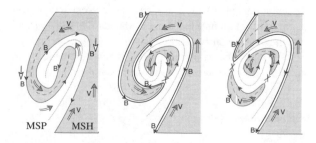

Figure 6. Sketch of the evolution of magnetic reconnection in a KH vortex at three successive times.

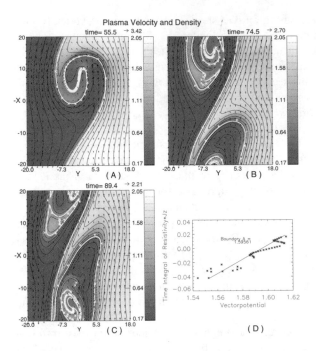

Figure 7. Evolution of a KH wave for three different times. The figures show plasma velocity (arrows), field lines projected into the x, y plane, and density(as grayscale). Panel (D) shows a test of the boundary location in terms of the integral electric field and the vectorpotential at the location of the plasma elements.

elements. Ideally the elements stay on the same field line if the magnetic field is frozen into the plasma flow. However, reconnection and numerical dissipation can break this condition.

Using the z component of the vector potential Ohm's law can be written as

$$\frac{dA_z}{dt} = \frac{\partial A_z}{\partial t} + \mathbf{V} \cdot \nabla A_z = -\eta j_z$$

with the resistivity η and the current density component j_z. Thus A_z for a plasma element is conserved if $\eta j_z = 0$ at all points of the trajectory. Magnetic field lines projected into the x, y plane are lines of constant A_z. The plasma elements are used (a) to determine the correct integration constant (correct gauge) for A_z and to ensure that numerical diffusion is sufficiently small. Figure 7d shows the time integral of ηj_z vs. the value of A_z for the fluid elements. The slope of the regression line in Figure 7 is very close to unity indicating that numerical diffusion is sufficiently small.

The mass entry can be expressed in terms of an average entry velocity through

$$v_{entry} = \frac{\Delta M}{\Delta t}\frac{1}{\rho_{sh}L_x}$$

where M is the mass in the magnetic islands, Δt is the simulation time, L_x is the wavelength, and ρ_{sh} is the magnetosheath density. For the case shown in Figure 7 the entry velocity is plotted in Figure 8. Mass entry starts when reconnection first occurs and reaches a maximum of about 1.4 km/s shortly after this time. Figure 9 shows the maximum entry velocity for all simulation cases.

Errors in these entry velocities can be caused by the choice of Δt and by inaccuracies of the island boundaries. Because of the exponential growth of the KH mode a reduction of the initial perturbation by a factor of 10 would only reduce the entry velocities by about 25%. The resistivity has a very minor influence on the results. For all presented cases the resistivity is chosen to depend on the current density in the form $\eta(j) = \kappa(j - J_c)S(j - J_c)$ with $\kappa = 0.02$, $J_c = 1.1J_0$, and S being the step function such that the resistivity is zero almost everywhere in the simulation domain. Even a choice of zero resistivity yields mostly unaltered results [*Nykyri and Otto*, 2001] because reconnection is forced by the thinning of the strong current sheets in the vortex motion of the KH instability.

In all cases mass transport occurs first from the high density to the low density side of the boundary. The magnetosheath arm of the vortex contains plasma which expands as it moves to the magnetospheric side (as seen in Figure7a). This stretching which is likely inertia driven leads to thinner magnetosheath vortex arms and stronger current sheets as demonstrated in Figure 4. Figure 9 shows that the initial field asymmetry has only a minor influence on the mass transport.

The results in Figure 9 can also be expressed as a diffusion coefficient (right scale in Figure 9). The typical diffusion coefficient is of the order of 10^9 m^{2}s^{-1}. The decreasing diffusion for increasing φ is caused by the increasing stabilization of the KH mode in particular in the nonlinear state for increasing tangential magnetic field. The limit of $\varphi = 0$ cannot be addressed by our method because there is no magnetic field in the x, y plane and thus no anti-parallel magnetic field developing. Assuming a plasma sheet density of 1

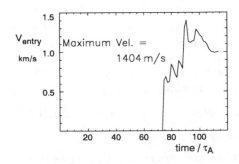

Figure 8. Average entry velocity for the case shown in Figure 7.

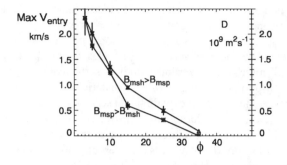

Figure 9. Maximum entry velocity for all simulation cases.

cm^{-3} the derived transport rates are sufficient to re-populate the plasma sheet with magnetosheath material in a few hours if plasma enters from the dusk and from the dawn side. This is consistent with the typical correlation time between solar wind and plasma sheet properties.

KH Modes and Reconnection in Three Dimensions

In three dimensions the two-dimensional constraints do not apply. In particular tearing modes and KH instabilities can have **k** vectors which are not anymore aligned. Figure 10 illustrates a configuration where the main shear velocity is along the y direction and the magnetic field is anti-parallel and aligned with the z direction.

The relation between the tearing mode and the KH for such configurations has been studied by *Pu et al.* [1990] and *Chen et al.* [1997]. For the chosen configuration both the tearing and the KH mode can operate because the main flow is perpendicular to the **k** vector of the fastest growing tearing mode and the magnetic field is perpendicular to the **k** vector of the fastest growing KH mode. For linear perturbations these modes grow independent of each other. Depending on boundary conditions and initial perturbations either mode can be dominant. Here we will focus on the case where the KH is the first mode to reach a nonlinear state.

The model is similar to the previously discussed 2D cases but uses a 3D code. The initial configuration uses

$$B_z = B_{z0} \tanh x$$
$$v_y = v_0 \tanh x$$

with the other field and velocity components equal to zero. Since we are not focussing on a particular application we use normalized units in which the magnetic field is measured by any given field strength B_0, mass density is normalized by ρ_0, length scales are measured in units of the current (shear) layer width L_0, velocity in units of the Alfvén speed v_A, and time in units of an Alfvén transit time $\tau_A = L_0/v_A$. The initial configuration uses a small perturbation to excite the tearing and KH modes.

Figure 11 illustrates the evolution of this system in terms of the location and deformation of the current layer. The KH mode assumes a nonlinear amplitude at about $t = 75\tau_A$. A fully developed KH vortex is formed shortly afterwards. The current layer has almost no distortion in the z direction until $t = 160\tau_A$. However, only a few Alfvén times later the flux surface of the current layer is heavily distorted indicating an extremely rapid growth of the tearing mode.

This is substantiated in Figure 12 which illustrates magnetic field vectors for times $t = 160\tau_A$, $165\tau_A$, and $170\tau_A$ in a cut through the KH vortex at $y = 7$. At $t = 160\tau_A$ seven parallel current layers are created by the vortex motion but are barely distorted at this time. At $t = 170\tau_A$ many tearing modes, in partly overlapping islands of considerable size are visible indicating a very fast growth of tearing modes in the late evolution.

The fast growth is caused by two effects. First the initial current sheet thins and intesifies when it is swept into the KH vortex. This leads to a corresponding increase in the effective growth rate of the tearing mode. At the same time the KH vortex creates locally many parallel current layers. It is known that the tearing mode in multiple current layers systems grows significantly faster in particular in its nonlinear evolution and the growth is strongly determined by the distance of the current layers [*Yan et al.*, 1994]. The growth of a tearing island in one current layer leads to a compression and enhanced reconnection in the neighboring current layers with a corresponding increase in the overall growth rate.

The many tearing islands lead to overlapping magnetic flux surfaces and meandering flux tubes. Figure 13 shows different views of a magnetic flux tube which is formed by reconnection by in the overlapping tearing islands. The flux tube shown originates from one side of the original current layer and meanders to the other side. This process is similar to magnetic percolation suggested as a reconnection process for the dayside magnetopause. The topology of open and closed magnetic flux within the KH vortex is continuously changing in time, and open and closed flux areas are interwoven on very small spatial scales. Both properties will al-

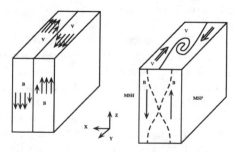

Figure 10. Configuration with shear flow and anti-parallel magnetic field in three dimensions.

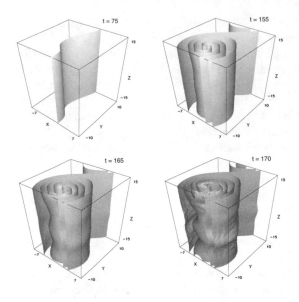

Figure 11. Illustration of the evolution of the initial current in several snapshots in time.

low for a very efficient mixing of plasma from the two sides of the original boundary.

DISCUSSION

The Kelvin Helmholtz mode is an ideal instability and as such not able to produce mass or magnetic flux transport across a plasma boundary. However there are various physical situations where the KH instability may well act not only as a catalyst but can be the key ingredient for reconnection to occur.

It is important to identify whether the KH mode indeed operates at the magnetospheric boundary because this determines the possible coupling processes between the magnetosheath and the magnetosphere. F1 and O1 made a strong case for such observational evidence. The properties of KH waves which are outlined in section 2 allow to clearly identify and distinguish KH waves in observations, certainly in cases where well structured KH vortices are encountered by satellites [*Fairfield et al.*, 2000; *Nikutowski et al.*, 2000; *Nykyri et al.*, 2001]. Among the many signatures are in particular unique modifications of the magnetic field at plasma boundaries (e.g. for outbound passes spikes in B_x and B_y of opposite polarity on the dusk side and of the same polarity on the dawn side [O1] accompanied by minima in B_z etc.) but also of the flow pattern, and plasma density and temperature. In addition to these properties the use of variance analysis, de Hoffmann-Teller frames, and the test of the Walen relation [*Nikutowski et al.*, 2000; *Nykyri et al.*, 2001] can provide further insight.

The KH mode has long been discussed as a source for viscous coupling at the magnetospheric boundary. Our re-

sult demonstrate that nonlinear KH modes cannot only trigger magnetic reconnection, they can render a largely parallel magnetic field into a configuration where reconnection is forced in the KH vortices. The plasma transport obtained in the two-dimensional study is sufficient to explain the transport of cold magnetosheath material into the magnetosphere and may thus explain the observations of the cold, dense plasma sheet during times of strongly northward IMF. Whether or not the nonlinear KH waves or cusp reconnection is the more likely process can presently not be decided. There is good evidence for both processes to occur during strongly northward field and it may well be that both cusp reconnection and the KH mode contribute to the cold, dense plasma sheet formation.

An interesting aspect for the KH mode is the stabilization of this mode by various conditions. First it is important to note that the local criteria are generally not applicable because the KH mode is a convective instability and as such is modifying the very environment where it operates and where a local stability analysis is considered. If the strucuture of the LLBL is caused be the KH mode a stability analysis of this very structure is ill posed.

There are various stabilizing physical aspects such as the magnetic field curvature, line tying to the ionosphere, formation of the LLBL due to other processes which may cause diffusion, and the magnetic field component aligned with the magnetosheath flow. Considering the detailed observations of F1 as confirmation for the KH mode has the important implication that the mode grows relatively fast. O1 have estimated that without any stabilizing influence the source region for the March 24, 1995, Geotail events are 10 to 16 R_E upstream, thus putting the source region close to the terminator. Any significant stabilization will lead to a lower growth rate and put the source region further upstream. It is difficult

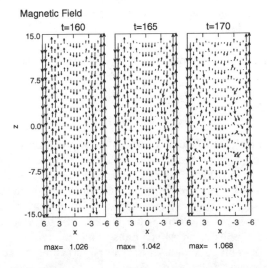

Figure 12. Cuts through the 3D system show magnetic field arrows for three times late in the simulation.

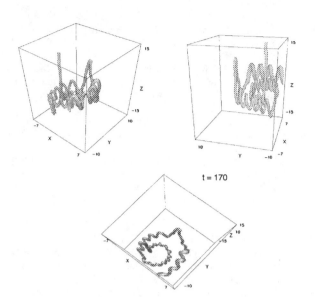

t = 170

Figure 13. Different views of a magnetic flux rope in the 3D configuration.

to quantify the effects of magnetic field curvature and ionospheric line tying. Field line curvature would localize the region where the KH mode operates and three-dimensional effects will be significant if the transverse scale becomes comparable to the wavelength (or the transverse coupling time through Alfvén waves comparable to the propagation time). The influence of ionospheric line tying is unresolved. This has been considered in some 2D studies through an integration of the drag along field-lines. However, the associated force is instantaneous while the real ionospheric feedback has a lag time.

It is interesting to consider competing processes which produce diffusion of the order of $D = 10^9 m^2 s^{-1}$ to generate the LLBL. The KH mode would grow significantly slower in a wide velocity reversal region and smaller tangential magnetic fields can entirely stabilize it. Thus we expect only small amplitudes for KH waves or no KH waves if other processes cause strong diffusion and generate the LLBL. Therefore the presence of nonlinear KH waves with only slightly reduced growth rates indicates that for such times the KH is likely to be the dominant process for generating the LLBL. This is also a noteworthy consideration for numerical simulations. Numerical simulations depend on the presence of some diffusion to avoid large numerical dispersion errors when wavelengths approach the grid scale. This numerical diffusion is necessary to obtain accurate physical results on scales larger than the grid scale. However, if the grid scale approaches the width of the boundary (or a reasonable fraction of it) numerical diffusion in global simulations can approach $D = 10^9 m^2 s^{-1}$ and can strongly damp the evolution of the KH mode. High resolution simulations can overcome

this problem and provide insight into the three-dimensional structure of the mode.

Our two-dimensional simulations show that reconnection in nonlinear KH can produce a mass transport consistent with observations. These results need to be examined more closely in three-dimensions. The basic process as we have documented here and in *Nykyri and Otto* [2001] will also operate in three dimensions. However, there is considerable additional physics which can play a role. In general one would expect that the magnetic topology is more complicated involving flux ropes in the KH vortices which can be entirely closed (i.e., with both ends in the ionosphere), open (connecting the magnetosphere and the ionosphere), and connected with both sides to the magnetosheath.

Our three-dimensional results illustrate this mixture and meandering of magnetic flux. They demonstrate several important additional new three-dimensional aspects of the interaction of KH and tearing modes. Nonlinear KH vortices not only generate new current layers, they also amplify the current density in pre-existing boundary layer currents. While the mixture of magnetic topology may imply that the effective mass transport is reduced (because a two-dimensionally captured magnetosheath filament would be expected to have mixed connection with only a portion entirely captured on closed magnetospheric flux) the possibility of three-dimensional orientations of **k** vectors for the tearing mode will certainly increase the reconnection and the amount of plasma affected by reconnection. The continuous change of this patchy magnetic topology should also strongly increase the anomalous diffusion in KH vortices suggested by *Fujimoto and Terasawa* [1994].

Much of the physics discussed in this paper addresses spatial structure. This includes the identification of KH modes at the magnetospheric boundaries, the physics of current layer formation within nonlinear flow vortices, magnetic reconnection within these vortices, and the resulting magnetic topology. We are looking forward to the observational results of multi-satellite missions such as Cluster where the identification of spatial structure and physical processes will be much easier than in single spacecraft observations.

Acknowledgments: The research was supported by the NASA SR&T Grant NAG-9457. K. Nykyri's work is supported by the Finnish Cultural Foundation and Finnish Academy. The computations were supported at the Arctic Region Supercomputer Center.

REFERENCES

Belmont, G., and G. Chanteur, Advances in magnetopause Kelvin-Helmholtz instability studies, *Phys. Scr.*, *40*, 124, 1989.

Birk, G. T., H. Wiechen, and A. Otto, Magnetic field amplification in M82 winds caused by Kelvin-Helmholtz modes, *Astrophys. J.*, *518*, 177, 1999.

Borovsky, J. E., M. F. Thomsen, and R. C. Elphic, The driving of the plasma sheet by the solar wind, *J. Geophys. Res.*, *103*, 17,617, 1998.

Chen, Q., A. Otto, and L. C. Lee, Tearing instability, Kelvin-Helmholtz instability, and magnetic reconnection, *J. Geophys. Res.*, *102*, 151, 1997.

Fairfield, D. H., R. P. Lepping, E. W. Hones, S. J. Bame, and J. R. Asbridge, Simultaneous measurements of magnetotail dynamics by IMP spacecraft, *J. Geophys. Res.*, *86*, 1396, 1981.

Fairfield, D. H., A. Otto, T. Mukai, S. Kokubun, R. P. Lepping, J. T. Steinberg, A. J. Lazarus, and T. Yamamoto, Geotail observations of the Kelvin-Helmholtz instability at the eqatorial magnetotail boundary for parallel northward fields, *J. Geophys. Res.*, *105*, 21,159–21,174, 2000.

Fujimoto, M., and T. Terasawa, Anomalous ion mixing within an MHD scale Kelvin-Helmholtz vortex, *J. Geophys. Res.*, *99*, 8601, 1994.

Fujimoto, M., and T. Terasawa, Anomalous ion mixing within an MHD scale Kelvin-Helmholtz vortex, 2, Effects of inhomogeneity, *J. Geophys. Res.*, *100*, 12,025, 1995.

Fujimoto, M., T. Terasawa, T. Mukai, Y. Saito, T. Yamamoto, and S. Kokubun, Plasma entry from the flanks of the near-Earth magnetotail: Geotail observations, *J. Geophys. Res.*, *103*, 4391, 1998.

Fujimoto, M., T. Mukai, A. Matsuoka, Y. Saito, H. Hayakawa, S. Kokubun, and R. P. Lepping, Multi-point observations of cold-dense plasma sheet and its relation with tail-llbl, *Adv. Space Res.*, *25*, 1607, 2000.

Keller, K. A., and R. L. Lysak, A two-dimensional simulation of kelvin-helmholtz instability with magnetic shear, *J. Geophys. Res.*, *104*, 25,097, 1999.

Lee, L. C., and J. V. Olson, Kelvin-helmholtz instability and the variation of geomagnetic pulsation activity, *Geophys. Res. Lett.*, *7*, 777, 1980.

Lennartsson, W., A scenario for solar wind penetration of the Earth's magnetic tail based on ion composition data from the ISEE 1 spacecraft, *J. Geophys. Res.*, *97*, 19,221, 1992.

Miura, A., Anomalous transport by magnetohydrodynamic Kelvin-Helmholtz instabilities in the solar wind magnetosphere interaction, *J. Geophys. Res.*, *89*, 801, 1984.

Miura, A., Simulation of the Kelvin-Helmholtz instability at the magnetospheric boundary, *J. Geophys. Res.*, *92*, 3195, 1987.

Miura, A., Kelvin-Helmholtz instability at the magnetospheric boundary: Dependence on the magnetosheath sonic Mach number, *J. Geophys. Res.*, *97*, 10,655, 1992.

Miura, A., Self-organization of the two-dimensional magnetohydrodynamic transverse Kelvin-Helmholtz instability, *J. Geophys. Res.*, *104*, 395, 1999.

Miura, A., and P. L. Pritchett, Nonlocal stability analysis of the MHD Kelvin-Helmholtz instability in a compressible plasma, *J. Geophys. Res.*, *87*, 7431, 1982.

Nikutowski, B., J. Büchner, A. Otto, L. M. Kistler, C. Mouikis, G. Haerendel, and W. Baumjohan, Equator-s observations of reconnection coupled to surface waves, *Adv. Space Res.*, *In Press*, 2000.

Nykyri, K., and A. Otto, Plasma transport at the magnetospheric boundary due to reconnection in kelvin-helmholtz vortices, *Geophys. Res. Lett.*, *28*, 3565, 2001.

Nykyri, K., A. Otto, J. Büchner, B. Nikutowski, W. Baumjohann, L. M. Kistler, and C. Mouikis, Equator-s observations of boundary signatures: Fte's or kelvin-helmholtz waves?, *AGU Monograph, Submitted contribution of the Chapman Conference on the LLBL*, 2001.

Otto, A., 3D resistive MHD computations of magnetospheric physics, *Comput. Phys. Commun.*, *59*, 185, 1990.

Otto, A., and D. H. Fairfield, Kelvin-Helmholtz instability at the magnetotail boundary: MHD simulation and comparison with Geotail observations, *J. Geophys. Res.*, *105*, 21,175, 2000.

Pu, Z. Y., M. Yan, and Z. X. Liu, Generation of vortex induced-tearing mode instability at the magnetopause, *J. Geophys. Res.*, *95*, 10,559, 1990.

Song, P., and C. T. Russell, Model for the formation of the low-latitude boundary layer for strongly northward interplanetary magnetic field, *J. Geophys. Res.*, *97*, 1411, 1992.

Terasawa, T., et al., Solar wind control of density and temperature in the near-earth plasma sheet: Wind/geotail collaboration, *Geophys. Res. Lett.*, *24*, 935, 1997.

Thomas, V. A., and D. Winske, Kinetic simulations of the Kelvin-Helmholtz instability at the magnetopause, *J. Geophys. Res.*, *98*, 11,425, 1993.

Wilber, M., and R. M. Winglee, Dawn-dusk asymmetries in the low-latitude boundary layer arising from the Kelvin-Helmholtz instability: A particle simulation, *J. Geophys. Res.*, *100*, 1883, 1995.

Wu, C. C., Kelvin-Helmholtz instability at the magnetopause boundary, *J. Geophys. Res.*, *91*, 3042, 1986.

Yan, M., A. Otto, D. Muzzel, and L. C. Lee, Tearing mode instability in a multiple current sheet system, *J. Geophys. Res.*, *99*, 8657, 1994.

A. Otto, K.Nykyri, Geophysical Institute, University of Alaska, Fairbanks, AK 99775-7320

Cluster Magnetic Field Observations of Magnetospheric Boundaries

T. S. Horbury, A. Balogh, M. W. Dunlop, P. J. Cargill, E. A. Lucek,
T. Oddy, P. Brown, and C. Carr

The Blackett Laboratory, Imperial College, London. U.K.

K.-H. Fornaçon

Institut für Geophysik und Meterologie, Technische Universität Braunschweig, Germany

E. Georgescu

MPI für Extraterrestrische Physik, Garching, Germany

Early Cluster magnetic field observations of boundaries and boundary layers in the near-Earth environment are reviewed. By combining data from four spacecraft, typically separated by $\sim$ 600 km, the scale, orientation and motion of structures can be deduced. Early results from the magnetic field investigation are reviewed, such as estimates of the orientation and motion of the magnetopause, including an inward velocity of 125 km s^{-1} in response to an interplanetary shock; evidence that crossings of the magnetopause are due to its inward and outward motion, rather than structures propagating along it; simultaneous flux transfer event signatures on both sides of the magnetopause; field-aligned currents and small scale structure in the mid-altitude cusp; variable motion of the high-altitude cusp; and the scale and orientation of mirror mode structures in the magnetosheath.

1. INTRODUCTION

Many questions regarding the physics of solar wind-magnetosphere coupling are related to boundary layers and their immediate environment. In particular, knowledge of the three dimensional structure and motion of these boundaries is essential to understand energy transfer through the system. For the first time, four-point measurements in the near-Earth environment at small and medium scales by the Cluster spacecraft allow us to estimate this structure and motion. In this paper, we review early magnetic field observations of a number of these boundaries and boundary layers: The magnetopause, cusp and magnetosheath.

While the Cluster formation does not pass through the low latitude boundary later (LLBL) itself, measurements of other regions, particularly the magnetopause and cusp, are highly relevant to studies of the LLBL.

1.1 Cluster: Mission, Spacecraft and Orbit

The Cluster mission is intended to study small scale plasma structures and their variations in space and time in

Earth's Low-Latitude Boundary Layer
Geophysical Monograph 133
Copyright 2003 by the American Geophysical Union
10.1029/133GM06

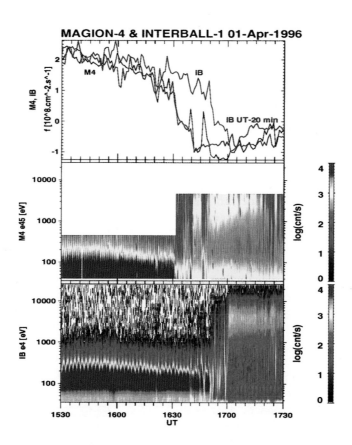

Plate 1. A comparison of IB1 and M4 observations of the low-latitude dayside magnetopause crossing. See text for the description.

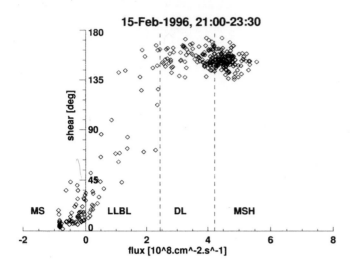

Figure 4. The ion flux measured by IB1 plotted as a function of magnetic shear.

from the sheath proper toward the magnetopause and further through the LLBL toward the plasma sheet. Distortions of the profile are caused by the variations of upstream parameters and occur simultaneously in measurements of both spacecraft. Two distinct changes of the upstream pressure are marked by dashed-dotted lines. The resulting distortions can be attributed to the scanning of the profile of a depletion layer suggested by *Seon et al.* [1999]. We will discuss this process later in this paper. The averaged profile lead to a determination of the depletion layer thickness about 1 R_E but the real thickness could be a little smaller. The thickness of the LLBL is about 0.5 R_E, i.e., similar to that determined in our previous example.

In principle, a profile like Figure 3 can occur without any real depletion as a consequence of averaging multiple crossings, if the magnetospheric intervals gradually increase. To show otherwise, we have plotted the magnetic shear as a function of the Faraday cup current used in Figure 4. One can clearly identify several plasma regimes: the sheath proper, depletion layer, where the density falls to 50% of the sheath density, the LLBL, where the overall change of magnetic field direction occurs (compare with Figures 1 and 2), and the plasma sheet characterized by a negative Faraday cup current caused by hot electrons.

A similarly good data ordering is exhibited in the plot of the electron density against the electron temperature presented in Figure 5. The data were taken during the whole interval depicted in Figure 2 with the 12 s time resolution. The scatter is enhanced by differing parallel and perpendicular temperatures. *Lockwood and Hapgood* [1997] suggested plotting the perpendicular temperature, which is not affected by bi-directional streams in the LLBL, but the temperature components cannot be reliably determined due to a highly

fluctuating magnetic field in this particular case. The ordering of the data in Figure 5 as well as in many similar cases [e.g., *Lockwood and Hapgood*, 1997; *Phan et al.*, 1997; *Vaisberg et al.*, 1998] suggests that the satellite crosses a region with continuously changing properties with the apparent fluctuations caused by oscillatory motion.

To show such an effect, we used a simple depletion layer model – a linear decrease by 50%. We plotted the magnetopause positions with respect to the M4 and IB1 Y_{GSE} coordinate in Figure 6. Until 2205 UT, the IB1 data were used as a sheath monitor and we calculated the magnetopause location from M4 data. Between 2205 and 2250 UT, we used a constant sheath density and calculated the magnetopause position from the IB1 data. The points, which lie along the spacecraft trajectories, are observed magnetopause crossings determined from the magnetic field rotation. Note a good match with the expected magnetopause locations.

Figure 6 demonstrates that even very complicated density profiles as that in Figure 2 can be explained by the motion of the magnetopause and adjacent layers. The source of a motion can be determined only by a careful analysis of the data but surface waves seem to be a prime candidate because observed upstream fluctuations are insufficient for a full explanation.

Another example is provided in Plate 2. The two top panels show the ion flux and magnetic field from IB1 during the magnetopause crossing on February 23, 1996. This crossing was observed at the same place and under similar upstream conditions as our previous example. The magnetic shear across the magnetopause was 145°. Three short excursions of IB1 from the sheath into magnetosphere at 1350, 1356, and 1400 UT can be probably attributed to upstream pressure fluctuations, as the computed distance of IB1 from the model magnetopause suggests. Consider the interval from 1402 to

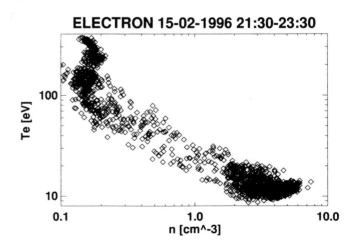

Figure 5. Electron temperature as a function of the electron density for the 2-hour interval involving all magnetopause boundary layers.

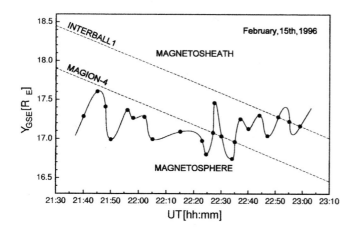

Figure 6. A simple model of scanning of the magnetopause and depletion layer. Points represent expected magnetopause positions with respect to IB1 and M4 Y-coordinates.

1422 UT. IB1 observed a smooth ion flux profile and the magnetic field is constant until 1414 UT. Then IB1 measured three current sheet crossings, seen as B_Y sign changes. Ion flux changes little in these crossings. On the other hand, M4 located 0.5 R_E inward of IB1, crossed all magnetopause layers, i.e., from the plasma sheet to sheath proper and vice versa. A careful examination of ion and electron energy spectra shown in last two panels (Plate 2) reveals that the ion fluxes above $\sim 7 \times 10^8$ cm^{-2} s^{-1} belong to the sheath proper and flux reduction to the value of $\sim 3.5 \times 10^8$ cm^{-2} s^{-1} occurs in a depletion layer. The fluxes below this level are observed in the LLBL.

This example shows that the oscillatory motion of magnetopause layers is intrinsic, and proceeds without external driving forces. The oscillations amplitude is less (probably significantly less) than the spacecraft separation. The depletion layer is 2000 km thick, as estimated from the computed magnetopause distance. The LLBL thickness thus estimated is 500-1000 km, i.e., just several ion gyroradii. An explanation of the fluctuations observed by M4 in terms of flux transfer events (FTEs) can be probably excluded because IB1 did not observe any signs of FTEs.

The LLBL is usually filled with tailward flowing plasma, although several cases of observations of sunward moving blobs of the low-energy plasma have been reported in the inner LLBL [e.g., *De Keyser et al.*, 2001]. However, the sunward flow of low-energy ions in the LLBL can be a significant LLBL feature under some circumstances as we document in Plate 3. The separation of IB1 and M4 was negligible (~ 120 km) during this event as revealed by the computed distance of the satellites from the magnetopause (4th panel).

The magnetopause was crossed at 0048 UT, and no PDL was seen. The crossing can be identified by a field rota-

tion (1st panel), by the appearance of high-energy electrons (3rd panel), and a switch between tailward streaming (5th panel) and sunward streaming (6th panel) ions. The behavior of electrons inferred from the Faraday cup is shown in the last panel. This Faraday cup is perpendicular to the satellite spin axis and thus one can see two features: rising mean value which corresponds to the increases in isotropic plasmaspheric electrons, and twice per spin observed spikes of streaming low-energy electrons. These counterstreaming beams indicate either closed field lines or continuous entry behind the satellite with reflection in the ionosphere. The ions are moving sunward during the whole interval of LLBL observations (a few exceptions will be discussed later). The v_X-component of the velocity (2nd panel) computed from the 3-D distribution is $\sim 50 - 100$ km/s. However, the energy spectrograms show that the sunward streaming low-energy (up to 5 keV) population is even faster than the sheath ions. The peak energy is about 1 keV in the LLBL, whereas it is about 700 eV in the sheath. Corresponding velocities are ~ 400 and ~ 330 km/s, respectively. A relatively low velocity obtained from 3-D distribution is affected by the presence of hot plasma sheet ions which are nearly stationary. These ions can be seen in the top edge of the E_{i0} spectrogram. The LLBL plasma is observed for ~ 50 minutes in this particular case but when we look at the satellite path with respect to the magnetopause (4th panel), we can see that the thickness of this layer is about 0.5 R_E.

In spite of high magnetic shear at the magnetopause, the sunward flow of LLBL ions cannot be explained by reconnection tailward of the spacecraft because the sheath velocity is supersonic at the spacecraft position and the flows would remain tailward. However, IMF B_Z was negligible (± 0.5 nT) during this event and thus the other magnetic field components become important for determination of the site where the antiparallel merging can occur.

4. DISCUSSION

The mechanisms leading to the formation of magnetopause layers for different IMF orientations are shown in Figures 7 and 8. If IMF is southward (Figure 7), subsolar reconnection can form the magnetopause current sheet as demonstrated by field line kinks. Inbound parts of the reconnected field lines belong to the LLBL, whereas outbound parts create "sheath transition layer" [*Russell*, 1995]. Plasma properties differ in these parts due to the different history. The LLBL part originally contained a hot, tenuous magnetospheric plasma and reconnection added an accelerated sheath population. A mixture of these populations exists inside the LLBL.

Sheath plasma is accelerated during reconnection, but an observer outside the subsolar point can not readily identify a velocity change. However, density is depleted inside the

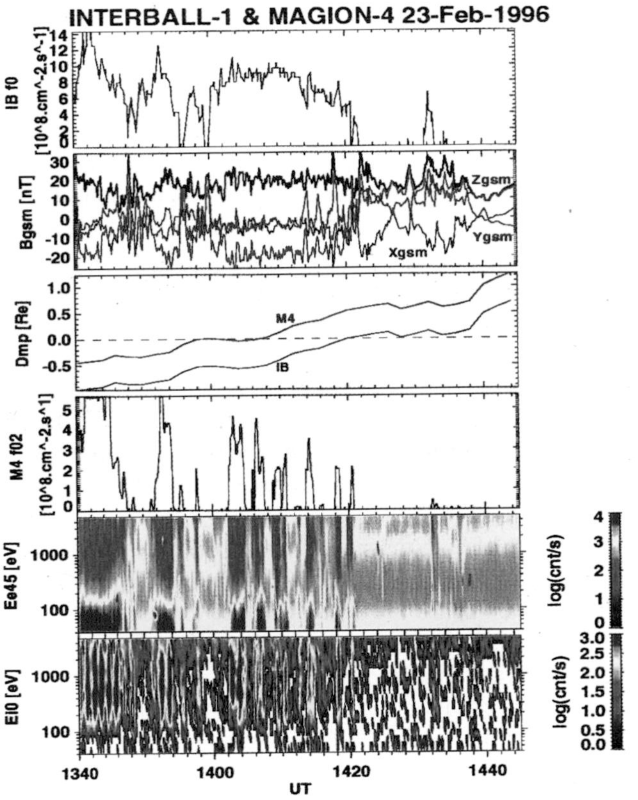

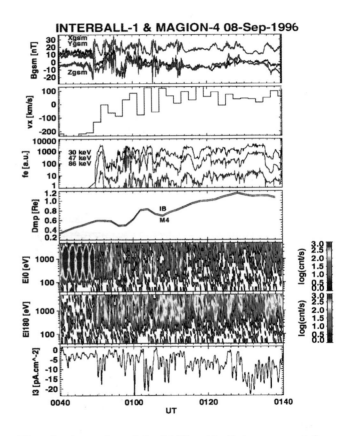

Plate 2. Two-point observations of a crossing of the low-latitude magnetopause. From top to bottom: the ion flux (IB1); magnetic field magnitude and components (IB1); the distance of IB1 and M4 from the model magnetopause; the ion flux (M4); electron (E_{e45}) and ion (E_{i0}) energy spectrograms measured by M4.

Plate 3. A crossing of the LLBL with distinct sunward flow. From top to bottom: components of the IB1 magnetic field; v_X-component of the ion velocity; fluxes of high-energy electrons; the distance of IB1 and M4 from the model magnetopause; energy spectra of tailward (E_{i0}) and sunward (E_{i180}) streaming ions; current density measured by a rotating Faraday cup.

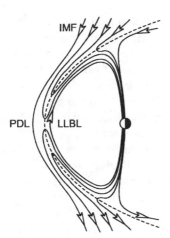

Figure 7. A sketch of the subsolar magnetopause for high magnetic shear.

layer because a part of the plasma was transmitted to the LLBL. For this reason, we denote this layer as a depletion layer (DL) in Figure 7.

Magnetospheric ions cannot follow the kink on the field lines because their gyroradius is too large, but the electrons can, and stream outward. In the LLBL, the electrons accelerated during reconnection move toward the ionosphere. Magnetic mirroring can produce counterstreaming beams. A similar feature cannot be seen in ion-energy spectra because the mirror time is comparable to the $\mathbf{E} \times \mathbf{B}$ drift time. The LLBL almost covers the frontside below the cusp in this scenario. But since all reconnected field lines eventually enter the mantle, it cannot explain the flank LLBL. Indeed, the flank LLBL is thinner vanishes during southward IMF [*Mitchell et al.*, 1987].

The situation for northward IMF is illustrated in Figure 8, as suggested by *Song and Russell* [1992] and *Le et al.* [1996]. Reconnection occurs poleward of the cusp (point 1) and newly reconnected field lines cover the dayside magnetopause. They may reconnect in the opposite hemisphere (point 2), closing a mixture of sheath and magnetospheric plasma inside the magnetosphere. Convection transfers the LLBL lines tailward creating the flank. However, as the convection is slow, the plasma parameters would evolve and one can expect that parameters of the flank LLBL created by this mechanism would differ from those in the subsolar region. Separately, magnetic field pile-up creates the PDL as described in the introduction. The low-latitude magnetopause represents a tangential discontinuity with no direct connection between the PDL and LLBL in this case.

The presence of the LLBL along the flanks is probably caused by reconnection near the cusp. The merging location is affected by all IMF components. The positive (negative) B_Y component shifts the site to the later (earlier) local times

in the northern hemisphere and newly opened field lines in the cusp region move to the flank instead of to the subsolar region. This situation is schematically depicted in Figure 9. The satellite is located tailward of the dawn terminator in the LLBL, just below the neutral sheet. The field line crossing the spacecraft closes through the dawn cusps. After merging, the injected ions move along the line from the northern to southern hemisphere, and move sunward at the satellite location. A small change of the LLBL magnetic field can cause apparent motion of the satellite from point 1 to point 2, where the ions will move tailward as expected in the LLBL. The continuous stream of ions seen in Plate 3 suggests that reconnection is steady and that the LLBL is open. When B_Y becomes positive, the process reverses. Reconnection at the dawn magnetopause occurs in the southern hemisphere and a satellite below the neutral line observes a tailward flow. Intervals of purely southward or northward IMF are rare and thus the process depicted in Figure 9 is probably most applicable to flank LLBL formation.

Plasma parameters plotted across the magnetopause usually exhibit large fluctuations. In many cases, these fluctuations can be connected with the presence of surface waves (Figure 6 and Plate 2). Such waves were reported many times [e.g., *Šafránková et al.*, 1997b; *Faruggia et al.*, 2000] and are attributed usually to the Kelvin-Helmholtz instability. Despite an extended discussion on whether the inner or the outer LLBL boundary is unstable, both often move together (Plate 2, and [*Šafránková et al.*, 1997b]). We believe that the waves cause a negative curvature of the local magnetopause, enhance diffusion [*Book and Sibeck*, 1995], and smooth the density profile. A similar suggestion can be found in *Phan et al.* [1997]. On the other hand, Plate 3 presents a case, where changes of all parameters are sharp and no waves are observed.

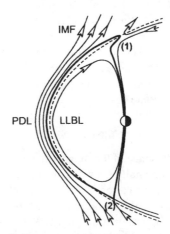

Figure 8. The formation of subsolar magnetopause layers when the magnetic shear is high.

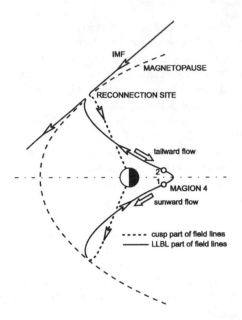

Figure 9. Schematic drawing of the formation of the flank LLBL.

MHD simulations [*Otto and Fairfield*, 2000] suggest that K-H instabilities can create sheath plasma blobs inside the LLBL. These blobs might explain the complex LLBL structure [e.g., *Vaisberg et al.*, 1998; *De Keyser et al.*, 2001]. However, such blobs would lead to the observation of negative density gradients inside the LLBL when the satellite closer to the magnetopause observes a lower density than that deeper in the LLBL. *Sibeck et al.* [2000] analyzed many IB1/M4 two-point observations and did not find any such case. Based on these observations, we conclude that such events are rare. The same observations tend to exclude impulsive penetration also.

We have determined the thickness of the LLBL in several cases and found it to be less than $0.5 R_E$. We think that reports of relatively thick LLBL [e.g., *Sauvaud et al.*, 1997; *Faruggia et al.*, 2000] are connected with the presence of surface waves with a large amplitude which apparently widen the region or with changes of upstream conditions causing the simultaneous motion of boundary layers and observing spacecraft, as in Plate 3.

5. CONCLUSION

Although the presence of the magnetopause boundary layers has been experimentally shown almost thirty years ago, the problem of their formation is still open. None of the proposed mechanisms alone can explain all observational features, although a combination might. We here summarized present knowledge and added new two-point observations to explain some peculiarities of boundary layers. Specifically we conclude that: (1) A layer of gradually decreased plasma density in front of the magnetopause is frequently present regardless of the IMF B_Z orientation and the value of the magnetic shear. (2) The thickness of this layer (from two-point observations) varies from 0.6 to 1 R_E. (3) The plasma depletion is not necessarily associated with magnetic field enhancements. (4) Field-aligned electron beams are often observed inside the depletion layer. (5) The depletion layer is formed by the magnetic field pile-up at the low-shear subsolar magnetopause and by reconnection when the magnetic shear is high. (6) Sheath plasma enters the dayside LLBL via subsolar reconnection when the IMF is southward, whereas reconnection tailward of the cusp can supply this region during intervals of northward IMF. (7) The thickness of the flank LLBL lies from 0.1 to 0.5 R_E, i.e., similar to the thickness of the dayside LLBL. (8) From our investigation, it follows that the most probable source of low-energy plasma in the flank LLBL is high-latitude reconnection near the cusp region. (9) The location of a reconnection site is determined by the direction of the sheath magnetic field. For a particular orientation, one flank is supplied from the southern hemisphere and the other from the northern hemisphere. (10) Diffusion is enhanced by surface waves on the magnetopause and can result in a smooth density profile of the whole sheath - plasma sheet interface. (11) The resulting profile exhibits a layer of gradually depressed density on the sheath side of the magnetopause, which extends through the LLBL. (12) Scanning of this profile can be the source of fluctuations in spacecraft observations near the magnetopause.

Acknowledgments. The authors thank P. T. Newell for his help in adjustment of this paper, R. Lepping for the WIND magnetic field and A. Lazarus for the WIND plasma data, and the INTERBALL team for the magnetic field data. The present work was supported by the Czech Grant Agency under Contract 205/00/1686, by the Charles University Grant Agency, Contracts 181 and 163, and by the Research Project, MSM 113200004. The Editor would like to thank the reviewer of this manuscript.

REFERENCES

Anderson, B. J., and S. A. Fuselier, Magnetic pulsation from 0.1 to 4.0 Hz and associated plasma properties in the Earth's subsolar magnetosheath and plasma depletion layer, *J. Geophys. Res.*, *98*, 1461, 1993.

Anderson, B. J., T.-D. Phan, and S. A. Fuselier, Relationship between plasma depletion and subsolar reconnection, *J. Geophys. Res.*, *102*, 9351, 1997.

Book, D. L., and D. G. Sibeck, Plasma transport through the magnetopause by turbulent interchange processes, *J. Geophys. Res.*, *100*, 9567-9573, 1995.

Cahill, L. J., and P. G. Amazeen, The boundary of geomagnetic field, *J. Geophys. Res.*, *68*, 1835, 1963.

De Keyser, J., F. Darrouzet, M. Roth, O. L. Vaisberg, N. Rybjeva, V. Smirnov, L. Avanov, Z. Němeček, and J. Šafránková, Transients at the dawn and dusk side magnetospheric boundary: Surface waves or isolated plasma blobs? *J. Geophys. Res.*, *106*, 25503-25516, 2001.

Eastman, T. E., E. W. Hones Jr., S. J. Bame, and J. R. Asbridge, The magnetospheric boundary layer: Site of plasma, momentum and energy transfer from the magnetosheath into the magnetosphere, *Geophys. Res. Lett.*, *3*, 685-688, 1976.

Eastman, T. E., and E. W. Hones Jr., Characteristics of the magnetospheric boundary layer and magnetopause layer as observed by IMP 6, *J. Geophys. Res.*, *84*, 2019, 1979.

Faruggia, C. J., et al., Coordinated Wind, Interball/tail, and ground observations of Kelvin-Helmholtz waves at the near-tail, equatorial magnetopause at dusk: January 11, 1997, *J. Geophys. Res.*, *105*, 7639, 2000.

Fuselier, S. A., D. M. Klumpar, E. G. Shelley, B. J. Anderson, and A. J. Coates, He^{2+} and H^+ dynamics in the subsolar magnetosheath and plasma depletion layer, *J. Geophys. Res.*, *96*, 21095, 1991.

Fuselier, S. A., B. J. Anderson, and T. G. Onsager, Particle signatures of magnetic topology at the magnetopause: AMPTE/CCE observations, *J. Geophys. Res.*, *100*, 11805, 1995.

Gosling, J. T., M. F. Thomsen, S. J. Bame, and C. T. Russell, Accelerated plasma flows at the near tail magnetopause, *J. Geophys. Res.*, *91*, 3029, 1986.

Hall, D. S., C. P. Chaloner, D. A. Bryant, D. A. Lepine, and V. P. Triakis, Electrons in the boundary layers near the dayside magnetopause, *J. Geophys. Res.*, *96*, 7869, 1991.

Klimov, S., et al., ASPI experiment: Measurements of fields and waves onboard the Interball-1 spacecraft, *Ann. Geophys.*, *15*, 514-527, 1997.

Kudela, K., M. Slivka, J. Rojko, and V. N. Lutsenko, The apparatus DOK-2 (project INTERBALL): Output data structure and modes of operation, *Tech. Rep. UEF-01-95*, pp.65, Institute of Experimental Physics, Slovak Academy of Science, Kosice, Slovakia, 1995.

Le, G., C. T. Russell, J. T. Gosling, and M. F. Thomsen, ISEE observations of low-latitude boundary layer for northward interplanetary magnetic field: Implications for cusp reconnection, *J. Geophys. Res.*, *101*, 27239, 1996.

Lockwood, M., and M. A. Hapgood, How the magnetopause transition parameter works, *Geophys. Res. Lett.*, *24*, 373-376, 1997.

Mitchell, D. G., F. Kutchko, D. J. Williams, T. E. Eastman, L. A. Frank, and C. T. Russell, An extended study of low-latitude boundary layer on the dawn and dusk flanks of the magnetosphere, *J. Geophys. Res.*, *92*, 7394, 1987.

Němeček, Z., A. Fedorov, J. Šafránková, and G. Zastenker, Structure of the low–latitude magnetopause: MAGION–4 observation, *Ann. Geophys.*, *15*, 553-561, 1997.

Otto, A., and D. H. Fairfield, Kelvin-Helmholtz instability at the magnetotail boundary: MHD simulation and comparison with Geotail observations, *J. Geophys. Res.*, *105*, 21175, 2000.

Paschmann, G., B. U. O. Sonnerup, I. Papamastorakis, N. Sckopke, G. Haerendel, S. J. Bame, J. R. Asbridge, J. T. Gosling, C. T. Russell, and R. C. Elphic, Plasma acceleration at the Earth's magnetopause: Evidence for magnetic reconnection, *Nature*, *282*, 243, 1979.

Paschmann, G., B. Sonnerup, I. Papamastorakis, W. Baumjohann, N. Sckopke, and H. Luchr, The magnetopause and boundary layer for small magnetic shear: Convection electric fields and reconnection, *Geophys. Res. Lett.*, *17*, 1829, 1990.

Paschmann, G., W. Baumjohann, N. Sckopke, T. D. Phan, and H. Luhr, Structure of the dayside magnetopause for low magnetic shear, *J. Geophys. Res.*, *98*, 13409, 1993.

Phan, T., G. Paschmann, W. Baumjohann, N. Sckopke, and H. Luhr, The magnetosheath region adjacent to the dayside magnetopause: AMPTE/IRM observations, *J. Geophys. Res.*, *99*, 121, 1994.

Phan, T. D., and G. Paschmann, Low-latitude dayside magnetopause and boundary layer for high magnetic shear, 1. Structure and motion, *J. Geophys. Res.*, *101*, 7801, 1996.

Phan, T. D., et al., Low-latitude dusk flank magnetosheath, magnetopause, and boundary layer for low magnetic shear: Wind observations, *J. Geophys. Res.*, *102*, 19883, 1997.

Russell, C. T., The structure of the magnetopause, *in Physics of the Magnetopause*, edited by P. Song, B. U. O. Sonnerup, and M. F. Thomsen, pp. 81-98, Geophysical Monograph 90, 1995.

Šafránková, J., G. Zastenker, Z. Němeček, A. Fedorov, M. Simerský, and L. Přech, Small scale observation of the magnetopause motion: Preliminary results of the INTERBALL project, *Ann. Geophys.*, *15*, 562, 1997a.

Šafránková, J., Z. Němeček, L. Přech, G. Zastenker, A. Fedorov, S. Romanov, J. Šimunek, and D. Sibeck, Two-point observation of magnetopause motion: INTERBALL project, *Adv. Space Res.*, *20*, 801, 1997b.

Sauvaud, J.-A., P. Koperski, T. Beutier, H. Barthe, C. Aoustin, J. J. Throcaven, J. Rouzaud, E. Penou, O. Vaisberg, and N. Borodkova, The INTERBALL- Tail ELECTRON experiment: Initial results on the low-latitude boundary layer of the dawn magnetosphere, *Ann. Geophys.*, *15*, 587-595, 1997.

Seon, J., S. M. Park, K. W. Min, L. A. Frank, W. R. Paterson, and K. W. Ogilvie, Observations of density fluctuations in Earth's magnetosheath with Geotail and Wind spacecraft, *Geophys. Res. Lett.*, *26*, 959, 1999.

Shue, J.-A., J. K. Chao, H. C. Fu, C. T. Russell, P. Song, K. K. Khurana, and H. J. Singer, A new functional form to study the solar wind control of the magnetopause size and shape, *J. Geophys. Res.*, *102*, 9497, 1997.

Sibeck, D. G., L. Přech, J. Šafránková, and Z. Němeček, Two-point measurements of the magnetopause: INTERBALL observations, *J. Geophys. Res.*, *105*, 237, 2000.

Song, P., and C. T. Russell, A model of the formation of the low-latitude boundary layer for strongly northward interplanetary magnetic field, *J. Geophys. Res.*, *97*, 1411, 1992.

Song, P., C. T. Russell, and M. F. Thomsen, Slow mode transition in the frontside magnetosheath, *J. Geophys. Res.*, *97*, 8295, 1992.

Southwood, B. U. O., and M. G. Kivelson, Magnetosheath flow near the magnetopause: Zwan-Wolf and Southwood-Kivelson theories reconciled, *Geophys. Res. Lett.*, *22*, 3275, 1995.

Vaisberg, O. L., V. N. Smirnov, L. A. Avanov, J. H. Waite, Jr., J. L. Burch, C. T. Russell, A. A. Skalsky, and D. L. Dempsey,

Observation of isolated structures of the low latitude boundary layer with the INTERBALL/Tail probe, *Geophys. Res. Lett.,* *25,* 4305-4308, 1998.

Wu, C.C., MHD flow past and obstacle: Large-scale flow in the magnetosheath, *Geophys. Res. Lett., 19,* 87, 1992.

Yermolaev, Y. I., A. O. Fedorov, O. L. Vaisberg, V. M. Balebanov, Y. A. Obod, R. Jimenez, J. Fleites, L. Liera, and A. N. Omelchenko, Ion distribution dynamics near the Earth's bow shock: First measurements with the 2-D ion energy spectrometer CORALL on the INTERBALL/Tail-probe satellite, *Ann. Geophys., 15,* 533, 1997.

Zwan, B. J., and R. A. Wolf, Depletion of the solar wind plasma near a planetary boundary, *J. Geophys. Res., 81,* 1636, 1976.

Z. Němeček, L. Přech, and J. Šafránková, Faculty of Mathematics and Physics, Charles University, V Holešovičkách 2, 180 00 Prague, Czech Republic. (zdenek.nemecek@ mff.cuni.cz)

J. A. Sauvaud, Centre d'Etude Spatiale des Rayonnements, BP4346, F-31029, Toulouse, France.

Enhanced Magnetospheric/Boundary Layer Plasma Flows Observed During Transient Magnetopause Crossings

K.-H. Kim,[1] N. Lin,[1] C. A. Cattell,[1] D.-H. Lee,[2] S. Kokubun,[3] T. Mukai,[4] and K. Tsuruda[4]

We observed enhanced plasma flows inside the magnetopause while the Geotail satellite briefly crossed the magnetopause. The enhanced flows were mainly in the MN plane of the LMN coordinates. Some of them showed a bipolar signature, i.e., inward flow before the outbound (from the magnetosphere to the magnetosheath) crossing and then outward flow after the inbound (from the magnetosheath to the magnetosphere) crossing, in the component normal to the nominal magnetopause. We found two different types of the bipolar flow: one is roughly symmetric with respect to the center of the event, that is, the peak amplitudes of the inward and outward flows are comparable, and the other is strongly asymmetric, that is, the outward flow speed is much larger than the inward flow speed. Using a simple qualitative model, we show that the symmetric bipolar flow is consistent with a vortical plasma motion from the $\mathbf{E} \times \mathbf{B}$ drift. The source of the electric field $\mathbf{E}$ may be attributed to briefly compressed magnetopause moving tailward, which is induced by a transient external (solar wind/foreshock) pressure pulse. In the asymmetric case, the strong outward flows were accompanied by a depressed magnetic field strength. This suggests that the strong outward plasma motion is associated with transient magnetospheric expansion driven by external pressure pulse variations.

1. INTRODUCTION

Transient magnetic field and plasma flow variations are commonly observed near the dayside magnetopause.

When transient events are observed with a prominent bipolar perturbation in the magnetic field component normal to the nominal magnetopause, they are referred to as flux transfer events (FTEs) and interpreted as transient and patchy reconnection at the dayside magnetopause [*Russell and Elphic*, 1978]. FTEs have been observed in the magnetosphere and/or in the magnetosheath. The magnetospheric FTEs are often accompanied by high-speed plasma flows and their plasma population is virtually identical to the low-latitude boundary layer (LLBL) plasma [e.g., *Paschmann et al.*, 1982; *Labelle et al.*, 1987; *Farrugia et al.*, 1988; *Elphic*, 1995]. Such high-speed flows have been usually interpreted as a reconnection-related phenomenon.

Recently, *Kim et al.* [2001a] (hereinafter referred to as Paper 1) observed quasi-periodic pulses of the low-latitude boundary layer (LLBL) plasma during southward IMF intervals. During the pulsed events, the

[1]School of Physics and Astronomy, University of Minnesota, Minneapolis

[2]Department of Astronomy and Space Science, Kyung Hee University, Kyunggi, Korea

[3]Department of Earth and Planetary Sciences, Tokyo Institute of Technology, Tokyo, Japan.

[4]Institute of Space and Astronautical Science, Sagamihara, Japan

Earth's Low-Latitude Boundary Layer
Geophysical Monograph 133
Copyright 2003 by the American Geophysical Union
10.1029/133GM08

LLBL plasma flows were strongly enhanced in the MN plane of the LMN coordinates [*Russell and Elphic*, 1978] and bipolar field and flow signatures were observed in the component normal to the nominal magnetopause. The bipolar flows were accompanied by sunward flows in the low density part of the LLBL (i.e., inner LLBL). The authors interpreted the observed bipolar/sunward plasma motions as the local plasma response to tailward moving flux tubes. That is, the enhanced (bipolar/sunward) flows for the pulsed events are not directly related to magnetic reconnection.

Observations of sunward flows inside the magnetopause for transient events have been reported by *Sibeck* [1992] and *Sibeck and Smith* [1992]. The events were simultaneously observed by the IRM and UKS satellites, separated by 0.06 R_E, on October 28, 1984. According to Sibeck and Smith, transient sunward flows in the outer magnetosphere and LLBL are not necessarily caused by reconnection but may indicate the tailward motion of a wavy magnetopause or an FTE flux tube on the magnetopause. One of the transient events, which was observed at ~1046 UT, in their studies exhibited an asymmetric plasma flow in the component normal to the nominal magnetopause with respect to the center of the event (that is, weak inward flow near the leading edge of the the event, tailward flow near the center of the event, and strong outward flow near the trailing edge of the event). *Sibeck* [1992] and *Sibeck and Smith* [1992] suggested that such flow motions would be expected during a brief entry into the magnetosheath caused by a passage of wavy magnetopause motion and that the strong outward flow on the trailing part of the event is associated with the magnetospheric expansion. However the authors did not provide conclusive evidence for magnetopause crossings during the event. Instead, many previous studies have interpreted the event as a magnetospheric FTE [*Rijnbeek et al.*, 1987; *Labelle et al.*, 1987; *Farrugia et al.*, 1988; *Lockwood et al.*, 1988; *Lockwood and Hapgood*, 1998]. It is still questioned what the satellites really observed.

The hypothesis for sunward flows provided by *Sibeck* [1992] and *Sibeck and Smith* [1992] has been recently confirmed by simultaneous ground and satellite observations [*Kim et al.*, 2001b] (hereinafter referred to as Paper 2). Paper 2 showed that transient sunward plasma flow inside the magnetopause can be detected while a wavy magnetopause was propagating tailward. Furthermore, the plasma flow motions during brief magnetosheath entries in Paper 2 were very similar to those for the asymmetric bipolar event in *Sibeck* [1992] and *Sibeck and Smith* [1992]; that is, weak inward flow before the outbound (from the magnetosphere to the magne-

tosheath) crossing, tailward flow in the magnetosheath, and strong outward flow after the inbound (from the magnetosheath to the magnetosphere) crossing.

The purpose of this study is to examine the plasma flow motions observed inside the magnetopause during transient magnetosheath entries. In order to avoid arguing whether a satellite briefly enters the magnetosheath or not, we only used the cases of high-shear magnetopause crossings (i.e., the southward magnetosheath magnetic field). We observed strongly enhanced plasma flows inside the magnetopause during transient magnetosheath entries. They were mainly in the MN plane of the LMN coordinates. We suggest that the enhanced flows were caused by magnetopause perturbations moving tailward rather than magnetic reconnection.

This paper is organized as follows. Section 2 briefly describes the data sets used in this study. Section 3 presents the observations. We compare our observations with previous studies and discuss plasma motions inside the magnetopause during brief magnetosheath entries in section 4. Section 5 gives the conclusions.

2. DATA SETS

The Geotail data used in this study were acquired with the fluxgate magnetometer [*Kokubun et al.*, 1994], the double-probe electric field detector [*Tsuruda et al.*, 1994], and the plasma experiment [*Mukai et al.*, 1994]. The magnetic and electric fields are measured at a rate of 16 samples/s and 32 samples/s, respectively, but in our study we use the spin (~3 s) averages of the field data. The electric field data were used to calculate the $\mathbf{E} \times \mathbf{B}$ drift velocity. On Geotail the electric field measurements are made in the spin plane. The component parallel to the spin axis (nearly along the GSE z direction) was calculated, assuming $\mathbf{E} \cdot \mathbf{B} = 0$, as $E_z = -(B_x/B_z) E_x - (B_y/B_z) E_y$. The calculation was made only when $|B_x|/|B_z|$ and $|B_y|/|B_z|$ were less than 2 to avoid magnifying errors when B_z is small. The plasma data are calculated from the low-energy particle (LEP) ion instrument. The plasma parameters are sampled at ~12 s. The LEP ion sensors usually change from the energy-per-charge analyzer (EA) to the solar wind ion analyzer (SW) as Geotail crosses the dayside magnetopause from the magnetosphere to the magnetosheath. However, the SW data can exhibit large errors near the dayside magnetopause because the plasma flow direction is largely deflected out of the field of view. Therefore, we used only the LEP-EA ion data in this study.

The Geotail field and flow data are presented in the boundary normal (LMN) coordinate system [*Russell*

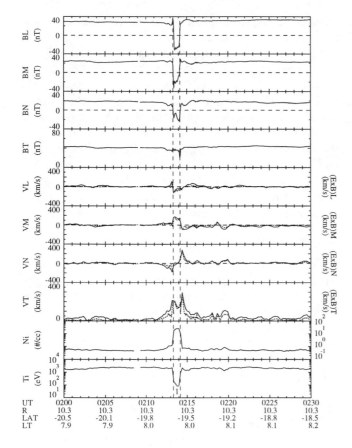

Figure 1. Geotail magnetic field and low-energy plasma (LEP) data on March 7, 1996. Components are plotted in LMN coordinates. The $\mathbf{E} \times \mathbf{B}$ drift velocities are plotted with the dotted lines in each component of the ion bulk flow. The vertical dashed lines indicate temporal magnetopause crossings.

and Elphic, 1978]. In the LMN system, the N direction points along the estimated outward normal to the model magnetopause at the spacecraft location, the L direction is a projection of the GSM z axis to the plane perpendicular to the N axis, and the M direction completes the right-handed coordinate system. To determine the direction normal to the magnetopause, the model magnetopause of *Roelof and Sibeck* [1993] for average values of IMF B_z and solar wind dynamic pressure was used, and it was deformed self-similarly to pass through the spacecraft position.

3. OBSERVATIONS

3.1. March 7, 1996

We examine first the Geotail outbound pass of March 7, 1996, through the morningside magnetopause. Figure 1 shows the time series of the magnetic data ($\sim$3 s) and LEP-EA plasma data ($\sim$12 s) in the model LMN

coordinates for the interval from 0200 to 0230 UT. The dotted lines in the ion bulk flow panels are the plasma drift velocities ($\sim$3 s) calculated from the magnetic and electric field assuming $\mathbf{V} = \mathbf{E} \times \mathbf{B}/|B|^2$. The calculated $\mathbf{E} \times \mathbf{B}$ velocities are in good agreement with the plasma flow velocities measured by LEP-EA.

The data in Figure 1 show that the satellite was mainly in the magnetosphere as evidenced by strongly northward magnetic field, low plasma density, and high temperature, and briefly encountered the magnetosheath for the interval of 0213:19-0214:10 UT (hereafter referred to as event 1). During the transient magnetosheath entry, the satellite observed an ion bulk flow of $\sim$150-200 km/s, ion temperature of $\sim$100-200 eV, and ion number density of $\sim$20-30 cm^{-3}.

Before the outbound crossing, the satellite observed a clear inward ($-V_N$) flow accompanied by northward ($+V_L$) and weak tailward ($+V_M$) flows. The peak position of the inward flow was consistent with that of the northward flow. After the inbound crossing, outward ($+V_N$) flow was observed and it was accompanied by sunward ($-V_M$) and weak southward ($-V_L$) flows. The inward (outward) motion in the V_N component during the outbound (inbound) crossing started (ended) in the magnetospheric region and their peak amplitudes were -180 km/s for the inward flow and $+280$ km/s for the outward flow, respectively. The bulk flow speed just before the magnetopause crossings was larger that the magnetosheath flow speed.

3.2. May 15, 1996

On May 15, 1996, Geotail also observed transient magnetosheath entries during an outbound pass through the morningside magnetopause. Figure 2 shows the magnetic field and plasma data for the interval from 1410 to 1500 UT using the same format as Figure 1. The missing LEP data in the magnetosheath are the periods measured by LEP-SW sensor. The missing $\mathbf{E} \times \mathbf{B}$ drift velocities are the periods where E_z is not calculated due to the constraint on the magnetic ratios described in section 2. As in the previous case, the satellite was mainly in the magnetosphere and observed transient magnetosheath entries during the intervals of $\sim$1426-1426:34 (event 2), $\sim$1448-1448:48 (event 3), $\sim$1449:44-1458:12 (event 4) UT.

During event 2, the satellite observed sharply increased magnetic field strength just before the outbound and after the inbound crossings. This suggests that the brief magnetosheath entry resulted from a transient magnetospheric compression. The plasma flow inside the magnetopause is predominantly in the normal component and showed inward ($-V_N$) motion before

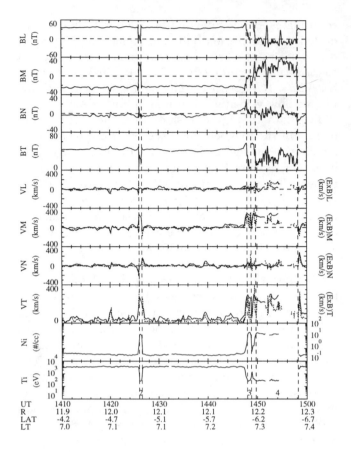

Figure 2. Same as Figure 1, but for May 15, 1996.

the outbound crossing and outward $(+V_N)$ motion after the inbound crossing. These signatures are very similar to those for event 1.

The inward/outward plasma motion in V_N was also observed at ~1420 UT in the magnetosphere without the magnetopause crossings (i.e., there were no significant plasma variations in density and temperature.). The event was accompanied by the sunward $(-V_M)$ and southward $(-V_L)$ flows. Such flows are consistent with the $\mathbf{E} \times \mathbf{B}$ drift as plotted in the figure. The southward flow may be due to the sunward rotation of the magnetic field (i.e., sharply duskward, $-B_M$, deflection).

In the B_N component, the bipolar (inward-then-outward) perturbation was clearly observed. Such a signature would be caused by a passage of reconnected flux tube (i.e., FTE) [*Russell and Elphic*, 1978]. However, this is not the only possible explanation. The bipolar B_N signature can be produced by a passage of a trough (i.e., transiently compressed magnetopause) [*Sibeck et al.*, 1989]. If the bipolar B_N event is caused by magnetopause motion, the bipolar field signature and the transient magnetosheath entry (event 2) can be at-

tributed to the same phenomena on the magnetopause, as reported in Paper 2.

The plasma flows for events 3 and 4 showed different signatures from those for events 1 and 2. That is, the inward $(-V_N)$ flow was weak or nearly zero, but the tailward $(+V_M)$ flow increased up to a value of magnetosheath flow speed before the outbound crossings at ~1448 UT for event 3 and 1449:44 UT for event 4. After the inbound crossing at 1458:12 UT, the weak sunward and strong outward flows were observed at 1458:30 UT in the magnetosphere. During the strong outward flow interval, the total magnetic field magnitude was smaller than that before the outbound crossing at 1449:44 UT.

3.3. April 29, 1996

The Geotail spacecraft was on an outbound pass on April 29, 1996, approaching the morningside magnetopause. During this orbit, Geotail observed a transient entry into the magnetosheath at ~1330 UT and multiple magnetopause crossings for the interval from ~1430 to ~1530 UT (data not shown). These crossings have been reported by *Kokubun et al.* [2000]. In this study, we will examine the relationship between the magnetic field and plasma flow variations inside the magnetopause for the transient magnetosheath entry at ~1330 UT (event 5) in detail.

The temporal outbound and inbound magnetopause crossings are plotted in Figure 3. As the satellite approached the magnetopause at 1329:42 UT, the magnetic field strength increased. The magnetic field maximum occurred just before the outbound crossing. After the inbound crossing at ~1332 UT, however, there was no significant magnetic field enhancement. The total magnetic field just after the inbound crossing was smaller than that just before the outbound crossing by ~20 nT.

The plasma flow velocities obtained from the LEP-EA sensor and the calculated $\mathbf{E} \times \mathbf{B}$ drift velocities are in good agreement with each other for the periods inside the magnetopause. Before the outbound crossing the flows were inward $(-V_N)$ and sunward $(-V_M)$. After the inbound crossing, they were strongly outward $(+V_N)$ and sunward $(-V_M)$. The outward flow reached a maximum value of ~430 km/s, which was faster than the magnetosheath flow by a factor of ~2.5, and it was observed near the earthward edge of LLBL and in magnetosphere. After the strong outward flow, the satellite observed inward $(-V_N)$ and tailward $(+V_M)$ plasma flow. There were no strongly enhanced plasma flows in V_L, that is, the plasma flow was predominantly in the MN plane. We note that the plasma flow motions inside

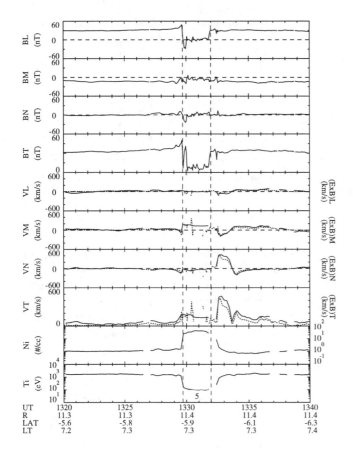

Figure 3. Same as Figure 1, but for April 29, 1996.

the magnetopause for event 5 are very similar to those of previous observations by the IRM satellite around 1046 UT on October 28, 1984 [*Sibeck*, 1992; *Sibeck and Smith*, 1992] and recent observations reported in Paper 2.

4. DISCUSSION

During events 1-5, the plasma flows observed inside the magnetopause were enhanced mainly in the MN plane of the LMN coordinates. They were in good agreement with the $\mathbf{E} \times \mathbf{B}$ drift velocity. These flow motions can be categorized by three types: (1) clear bipolar (inward/outward) flow in V_N (events 1 and 2), (2) strong outward flow accompanied by sunward flow after the inbound crossing (events 4 and 5), and (3) tailward flow accompanied by weak V_N flow before the outbound and after the inbound crossings (event 3).

In the first case, the negative and positive V_N peak amplitudes are roughly comparable. Such a bipolar V_N signature was also observed in the outer magnetosphere without transient magnetopause crossings and it was accompanied by a sunward flow (see Figure 2). Very

similar flow signatures have been reported in Paper 1. In that study bipolar (inward/outward) V_N perturbations accompanied by sunward flows were observed in the inner LLBL and attributed to local plasma response to the passage of flux tubes moving tailward, that is, the flow motions were not directly related to magnetic reconnection.

We can expect similar flow motions, a bipolar V_N flow accompanied by a sunward flow, inside the magnetopause when a perturbed magnetopause (i.e., a trough in the averaged magnetopause position), which can be produced by the magnetopause response to external pressure pulse, passes over a satellite. Figure 4, which is a modification of Figure 8 in Paper 1, illustrates a cut through a small portion of the equatorial magnetopause cross section on the morningside when a trough propagates tailward along the magnetopause. We assume that there is no reconnection on the magnetopause, that the trough speed on the magnetopause is comparable

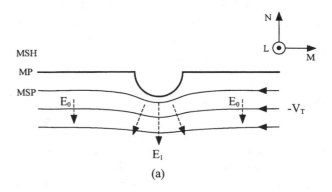

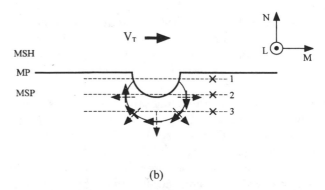

Figure 4. Schematic illustration of the magnetospheric plasma flows around the perturbed magnetopause. (a) Streamlines of the magnetospheric plasma flows in the rest frame of the perturbed magnetopause. (b) Plasma motions around the perturbed magnetopause moving tailward. Crosses indicate the satellite positions. View is from the north pole in the morningside. (Modified from Figure 8 of *Kim et al.* [2001a].) See the text for descriptions.

Table 1. Total Magnetic Field (B_T) and Plasma Density (N) at the Peaks of the Inward (i) and Outward (o) V_N Flows

Event	V_{Ni} (km/s)	V_{No} (km/s)	B_{Ti} (nT)	B_{To} (nT)	N_i (cm^{-3})	N_o (cm^{-3})	N_{MSP} (cm^{-3})	N_{MSH} (cm^{-3})
1	−180	+279	35.0	38.3	1.2	0.7	0.5-0.7	20-30
2	−135	+157	47.4	44.5	1.5	0.4	0.3-0.4	15-20
4	−12	+275	55.4	43.4	1.6	0.3	0.3-0.4	15-20
5	−149	+432	46.2	31.2	1.5	1.5	1.0-1.5	30-40

N_{MSP} and N_{MSH} indicate the density in the magnetosphere and magnetosheath, respectively. The V_{Ni}, B_{Ti}, and N_i for event 4 are the values at the peak of the bulk flow speed just before the outbound crossing because the inward flow was not significant.

to the magnetosheath flow speed, and that the magnetospheric plasma flow is stagnant. We also assume that the magnetic field in the magnetosphere is constant and directed northward (i.e., $+B_L$), that the plasma in the magnetosphere is incompressible and impenetrable about the magnetopause, and that the LLBL is absent inside the magnetopause.

We follow the same interpretation for Figure 4 in our study as in Paper 1, but the perturbed magnetopause is due to external pressure pulse. Figure 4a shows the magnetospheric plasma flow in the rest frame of the trough propagating tailward with a speed of $\mathbf{V}_T$. The curved lines indicate the streamlines. The streamlines far away from the perturbed magnetopause are undisturbed and therefore parallel to the nominal magnetopause. As the plasma flow approaches the trough, the streamlines go around it. Since the amount of flow per unit time across any section of two streamlines is the same, the flow velocities increase in the region around the trough. Therefore, the electric field $\mathbf{E}_1$, caused by $-\mathbf{V} \times \mathbf{B}$, on the perturbed streamlines is larger than $\mathbf{E}_0$ in the unperturbed region. The field $\mathbf{E}_0$ is removed by transforming the observations into the rest frame of the magnetospheric plasma. Thus only the perturbed field $\mathbf{E}_P$ (the dashed arrows), which is produced by perturbed magnetopause moving tailward, remains around the moving trough, and vortical motion can be expected from the $\mathbf{E}_P \times \mathbf{B}$ drift velocity (the solid arrows) as plotted in Figure 4b.

If a satellite is on trajectory 3, a bipolar (inward/outward) V_N flow accompanied by a sunward $(-V_M)$ flow would be observed as a trough passes over the satellite (e.g., the event at ~1420 UT on May 15, 1996, see Figure 2). On trajectory 2, the satellite may observe the inward (outward) flow inside the magnetopause just before (after) the outbound (inbound) crossing (e.g., events 1 and 2). If a satellite is on trajectory 1 (that is, just inside the nominal magnetopause), the electric fields in the regions just before the outbound

and after the inbound crossings would have the positive E_N component because the flow speed in those regions is smaller than $-\mathbf{V}_T$. Thus tailward flow $(+V_M)$ will be observed inside the magnetopause before the outbound and after the inbound crossings (e.g., event 3).

As a trough passes, the magnetospheric magnetic field under the trough will be compressed [Sibeck, 1990] and it can also be responsible for the flow enhancement around the trough. This implies that the total electric field around the trough in Figure 4a, which is caused only by an incompressible plasma flow, should increase. Consequently, the electric fields predicted on trajectory 2 in Figure 4b rotate inward (that is, inward/tailward before the crossing and inward/sunward after the crossing) and the $\mathbf{E} \times \mathbf{B}$ drift velocity has the sunward $(-V_M)$ component before the outbound and after the inbound crossings. This would explain why the sunward motion of the $\mathbf{E} \times \mathbf{B}$ was observed inside the magnetopause during events 1 and 2.

In the above model, we assumed the absence of the LLBL inside the magnetopause. However, the LLBL lies just inside all regions of the equatorial magnetopause [e.g., Eastman et al., 1976]. Previous studies by Fujimoto et al. [1996] and Phan et al. [1997] showed that the plasma flow speed in the LLBL is smaller than that in the adjacent magnetosheath and gradually decreases from the outer (high density) part of the LLBL to the inner (low density) part. The authors also reported that the flow is nearly aligned with the magnetosheath flow (i.e., tailward) throughout most of the LLBL, but in the inner part of the LLBL the flow is either stagnant or sunward. As noted above, the inward and outward flows during events 1 and 2 started and ended in the magnetosphere, respectively, and their peaks were observed near the earthward edge of the LLBL (or inner LLBL) as listed in Table 1. These observations suggest that the plasma motions in V_N are continuous in the magnetosphere and inner LLBL during the transient magnetosheath entries. In other words,

4. SUMMARY AND CONCLUSION

(1) During IMF southward, a layer-like energetic ions have been observed in the high latitude boundary region adjacent the magnetopause outside the magnetosphere (in the magnetosheath). The energetic particles in this region are highly anisotropic and exhibit a clear anti-sunward flow although geomagnetic activity is very quiet.

(2) For a IMF northward case, a layer-like energetic ions have been observed in the high latitude boundary region, however, this layer is located inside the magnetosphere. The energetic particles in this region are also highly anisotropic, but exhibit a clear sunward flow or bi-direction flow (sunward and antisunward) meanwhile geomagnetic activity is also very quiet.

(3) The observations for both IMF northward and southward case can be explained in a framework of convection models – dayside reconnection occurred in the subsolar region during a southward IMF B_z; high latitude reconnection happened during IMF northward. However, the reconnection processes are important simply because they either open a path for energetic magnetospheric ions to escape into the magnetosheath during southward IMF, or provide reverse convection geometry during northward IMF. The observed energetic particles for both cases in the high latitude boundary region may be provided by the tail plasma sheet particles because of a minimum magnetic field existing off the equator in the high latitude region of the magnetosphere (e.g. Tsyganenko magnetic field model). In this way, particles initially mirroring near the equator are expelled from low latitudes and subsequently swept into the boundary layer at high latitudes.

*Acknowledgments.*We are indebted to A. Balogh for provision of magnetic field data.

REFERENCES

Anderson, K. A., H. K. Harris, and R. J. Paoli, Energetic electron fluxes in and beyond the Earth's outer magnetosphere, *J. Geophys. Res.*, *70*, 1039–1050, 1965.

Antonova, A. E., High-latitude particle traps and related phenomena, *Radiat. Meas.*, *26(3)*, 409–411, 1996.

Baker, D. N., and E. C. Stone, The magnetopause electron layer along the distant magnetotail, *Geophys. Res. Lett.*, *4*, 133–136, 1977.

Chen, J., T. A. Fritz, R. B. Sheldon, H. E. Spence, W. N. Spjeldvik, J. F. Fennell, and S. Livi, A new, temporarily confined population in the polar cap during the August 27, 1996 geomagnetic field distortion period, *Geophys. Res. Lett.*, *24*, 1447–1450, 1997.

Chen, J., et al., Cusp energetic particle events: Implications for a major acceleration region of the magnetosphere, *J. Geophys. Res.*, *103*, 67–78, 1998.

Cowley, S. W. H., The causes of convection in the Earth's magnetosphere: A review of developments during the IMS, *Rev. Geophys.*, *20*, 531–565, 1982.

Crooker, N. U., Reverse convection, *J. Geophys. Res.*, *97*, 19,363–19.372, 1992.

Daly, P., Remote sensing of energetic particle boundaries, *Geophys. Res. Lett.*, *9*, 1329–1332, 1982.

Daly, P., and E. Keppler, Remote sensing of a flux transfer event with energetic particles, *J. Geophys. Res.*, *88*, 3971–3980, 1983.

Delcourt, D. C., and J.-A. Sauvaud, Populating of cusp and boundary layers by energetic (hundreds of kev) equatorial particles, *J. Geophys. Res.*, *104*, 22,635–22,648, 1999.

Dungey, J. W., Interplanetary magnetic field and the auroral zones, *Phys. Rev. Letters*, *6*, 47–48, 1961.

Dunlop, M. W., and T. I. Woodward, Multi spacecraft discontinuity analysis, in *Analysis Methods for Multi Spacecraft Data*, edited by G. Paschmann and P. W. Daly, pp. 271–305, ESA, Bern, Switzerland, 1998.

Fritz, T. A., The role of the cusp as a source for magnetospheric particles: A new paradigm, in *Proc. Cluster-II workshop on Multiscale/Multipoint Plasma Measurements*, pp. 203–209, Eur. Space Agency Spec. Publ., SP-449, ESA, 2000.

Fritz, T. A., The cusp as a source of magnetospheric energetic particles, currents, and electric fields: new paradigm, *Space Sci. Rev.*, *95*, 469–488, 2001.

Fritz, T. A., and S. C. Fahnenstiel, High temporal resolution energetic particle sounding at the magnetopause on november 8, 1977, using ISEE 2, *J. Geophys. Res.*, *87*, 2125–2131, 1982.

Fuselier, S. A., D. M. Klumper, W. K. Peterson, and E. G. Shelley, Direct injection of ionospheric O^+ into the dayside low latitude boundary layer, *Geophys. Res. Lett.*, *16*, 1121–1124, 1989.

Haerendel, G., On the 3-dimensional structure of plasmoids, *J. Geophys. Res.*, *83*, 3195–, 1978.

Haskell, G. P., Anisotropic fluxes of energetic particles in the outer magnetosphere, *Planet. Space Sci.*, *74*, 1740–1748, 1969.

Hones, E. W., S. I. Akasofu, S. J. Bame, and S. Singer, Outflow of plasma from the magnetotail into magnetosheath, *J. Geophys. Res.*, *77*, 6688–6695, 1972.

Lockwood, M., and M. F. Smith, The variation of reconnection rate at the dayside magnetopause and cusp ion precipitation, *J. Geophys. Res.*, *97*, 14,841–14,847, 1992.

Lundin, R., plasma composition and flow characteristic in the magnetospheric boundary layers connectedd to the polar cusp, in *The polar Cusp*, edited by J. A. Holtet and A. Egeland, pp. 9–32, Kluwer Academic Publishers, Dordrecht, the Netherlands, 1985.

Meng, C. I., and K. A. Anderson, A layer of energetic electrons (> 40kev) near the magnetopause, *J. Geophys. Res.*, *75*, 1827–1836, 1970.

Meng, C. I., A. T. Y. Lui, S. M. Krimigis, S. Ismail, and D. J. Williams, Spatial distribution of energetic particles in the distant magnetotail, *J. Geophys. Res.*, *86*, 5682–5700, 1981.

Moen, J., et al., Cluster boundary-layer measurements and optical observations at magnetically conjugate sites, *Ann. Geophys.*, *19*, in press, 2001.

Newell, P. T., and C. I. Meng, Cusp width and b_z: Observations and a conceptual model, *J. Geophys. Res.*, *92*, 13,673–13,678, 1987.

Paschmann, G., G. Haerendel, N. Sckopke, and H. Rosenbauer, Plasma and field characteristics of the distant polar cusp near local noon: The entry layer, *J. Geophys. Res.*, *81*, 2883–2899, 1976.

Roederer, J. G., On the adiabatic motion of the energetic particles in a model magnetosphere, *J. Geophys. Res.*, *72*, 981–992, 1967.

Rosenbauser, H., A boundary layer model for magnetospheric substorms, *J. Geophys. Res.*, *80*, 2723–, 1975.

Scholer, M., F. M. Ipavich, G. Gloeckler, D. Hovestadt, and B. Klecker, Leakage of the magnetospheric ions into the magnetosheath along reconnected field lines at the dayside magnetopause, *J. Geophys. Res.*, *86*, 1299–1304, 1981.

Sckopke, N., G. Paschmann, H. Resenbauer, and D. H. Fairfield, Influence of the interplanetary magnetic field on the occurrence and the thickness of the plasma mantle, *J. Geophys. Res.*, *81*, 2687–2691, 1976.

Shabansky, V. P., Magnetospheric process and related geophysical phenomena, *Space Sci. Rev.*, *8*, 366–454, 1968.

Sheldon, R. B., H. E. Spence, J. D. Sullivan, T. A. Fritz, and J. Chen, The discovery of trapped energetic electrons in the outer cusp, *Geophys. Res. Lett.*, *25*, 1825–1828, 1998.

Sibeck, D. G., R. W. McEntire, A. T. Y. Lui, R. E. Lopez, S. M. Krimigis, R. B. Decker, L. J. Zanetti, and T. A. Potemra, Energetic magnetospheric ions at the dayside magnetopause: Leakage or merging?, *J. Geophys. Res.*, *92*, 12,097–12,114, 1987.

Sibeck, D. G., et al., Plasma transfer processes at the magnetopause, in *Magnetospheric Plasma Sources and Losses*, edited by B. Hultqvist and M. Oieroset, pp. 207–283, Kluwer Acade., Norwell, Mass., 1999.

Song, P., and C. T. Russell, Model of the formation of low-latitude boundary layer for strongly northward interplanetary magnetic field, *J. Geophys. Res.*, *97*, 1411–, 1992.

Trattner, K. J., S. A. Fuselier, W. K. Peterson, S. W. Chang, R. Friedel, and M. R. Aellig, Origins of energetic ions in the cusp, *J. Geophys. Res.*, *106*, 5967–5976, 2001.

West, H. I., and R. M. Buck, Observations of > 100 kev protons in the Earth's magnetosheath, *J. Geophys. Res.*, *81*, 569–584, 1976.

Wilken, B., D. N. Baker, P. R. Higbie, T. A. Fritz, W. P. Olson, and K. A. Pfitzer, Magnetospheric configuration and energetic particle effects associated with a SSC: A case study of the CDAW event on March 22, 1979, *J. Geophys. Res.*, *91*, 1459–1473, 1986.

Wilken, B., et al., Rapid: The imaging energetic particle spectrometer on cluster, *Space Sci. Rev.*, *79*, 399–473, 1997.

Williams, D. J., Magnetopause characteristics inferred from three dimensional energetic particle distributions, *J. Geophys. Res.*, *84*, 101–104, 1979.

Yamauchi, M., and R. Lundin, Physical signatures of magnetospheric boundary layer processes, in *Physical Signatures of Magnetospheric Boundary Layer Processeses*, edited by J. A. Holtet and A. Egeland, pp. 99–109, Kluwer Academic Publishers, Dordrecht, the Netherlands, 1998.

Zong, Q.-G., and B. Wilken, Layered structure of energetic oxygen ions in the magnetosheath, *Geophys. Res. Lett.*, *25*, 4121–4124, 1998.

Zong, Q.-G., and B. Wilken, Bursty energetic oxygen events in dayside magnetosheath: Geotail observation, *Geophys. Res. Lett.*, *26*, 3349–3352, 1999.

Zong, Q.-G., B. Wilken, J. Woch, G. Reeves, T. Doke, and T. Yamamoto, Energetic particle bursts in the near-Earth magnetosheath during a storm, *Phys. Chem. Earth*, *24*, 293–298, 1999.

Zong, Q.-G., B. Wilken, S. Y. Fu, and Z. Y. Pu, Energetic oxygen ions sounding the magnetopause, in *Multiscale/ Multipoint Plasma Measurements*, pp. 379–385, Eur. Space Agency Spec. Publ., SP - 449, ESA, 2000.

Zong, Q.-G., B. Wilken, S.-Y. Fu, T. A. Fritz, Z.-Y. Pu, N. Hasebe, and D. J. Williams, Ring current oxygen ions in the magnetosheath caused by magnetic storm, *J. Geophys. Res.*, *106*, 25,541–25,556, 2001.

T. A. Fritz and Q.-G. Zong, Center for Space Physics, Boston University, Commonwealth Avenue 725, Boston, MA 02215, USA. (e-mail: fritz@bu.edu, zong@bu.edu)

P. Daly, B. Wilken, Max-Planck-Institut für Aeronomie, D-37191, Katlenburg-Lindau, Germany. (e-mail: daly@limpi.mpg.de; wilken@linmpi.mpg.de)

Low-latitude Boundary Layer Formation by Magnetic Reconnection

T. G. Onsager

NOAA Space Environment Center, Boulder, Colorado

J. D. Scudder

Department of Physics and Astronomy, University of Iowa, Iowa City, Iowa

Magnetic reconnection at the dayside magnetopause allows the direct transport of plasma into and out of the magnetosphere along the interconnected magnetic field, thereby forming boundary layers both inside and outside the magnetopause. The low-latitude portion of this boundary layer is referred to as the low-latitude boundary layer (LLBL) inside the magnetopause and as the magnetosheath boundary layer (MSBL) outside the magnetopause. As the intermixing plasma encounters the magnetopause current layer, it can be substantially modified through particle acceleration, heating, and partial reflection and transmission. As a result of these effects, the plasma in the LLBL is characterized by densities and temperatures intermediate between those in the neighboring magnetosheath and magnetosphere. The LLBL formation is quite different under southward and northward interplanetary magnetic field (IMF) conditions. When the IMF is southward, the LLBL forms directly at low latitudes where reconnection is occurring. During northward IMF, the LLBL forms through the remote occurrence of reconnection at high latitudes. In this paper, we present direct observations of the plasma heating and intermixing at high latitudes where reconnection is occurring under northward IMF. We demonstrate that the local entry and heating of the magnetosheath plasma across the high-shear, high-latitude magnetopause is sufficient to account for the observed LLBL properties at low latitudes.

1. INTRODUCTION

The low-latitude boundary layer (LLBL), as an extension of the magnetopause, is driven by reconnection and continuously evolves as reconnection changes in response to the changing interplanetary magnetic field (IMF).

Earth's Low-Latitude Boundary Layer
Geophysical Monograph 133
Copyright 2003 by the American Geophysical Union
10.1029/133GM11

Through reconnection and the subsequent evolution of the interconnected field lines, magnetosheath plasma originating outside the magnetopause and magnetospheric and ionospheric plasma originating inside the magnetopause are able to intermix forming boundary layers both inside and outside the magnetopause. In addition, plasma heating and acceleration occurs at the magnetopause current layer that modifies the boundary layer plasma relative to a simple mixture of the populations inside and outside the magnetopause. Although not always clearly discernible in spacecraft data, the magnetopause is usually identified as the location where the major rotation in the magnetic field occurs between the magnetosheath and magnetospheric orientations. The boundary layer outside the magnetopause is typically referred to as the magnetosheath boundary layer (MSBL) [e.g., *Cowley,* 1992], while the boundary layer inside the magnetopause is referred to as the low-latitude boundary layer.

The LLBL is identified as a region immediately inside the magnetopause with densities and temperatures that are intermediate between those in the neighboring magnetosheath and magnetosphere. As described in more detail below, the plasma properties in the LLBL result primarily from four main processes: the intermixing of plasma across the magnetopause, plasma acceleration at the magnetopause, plasma bulk heating at the magnetopause, and the velocity-space accessibility of the intermixing plasma away from the magnetopause. For a recent review of research on the magnetopause and LLBL, see *Sibeck et al.* [1999].

Although there is considerable uncertainty and probably much variability in precisely where magnetic reconnection operates on the magnetopause as a function of IMF clock angle, it predominantly occurs at low latitudes (equatorward of the cusps) under southward IMF conditions [e.g., *Gosling et al.,* 1990a] and at high latitudes (poleward of the cusps) for northward IMF [e.g., *Crooker,* 1992]. In the case of southward IMF reconnection, the LLBL forms directly at low latitudes, close to where reconnection is occurring. In the case of northward IMF, which is the primary focus of this paper, the LLBL forms remotely through the draping of field lines over the low latitude magnetopause that have reconnected at high latitudes.

When reconnection occurs under any orientation of the IMF and the magnetosheath and magnetospheric magnetic fields become interconnected, the plasmas on both sides of the magnetopause begin to intermix. Particles that then cross the magnetopause current layer typically undergo some energization. This energization can occur either as a coherent acceleration, with little change in the temperature of the intermixing plasma components, or as a bulk heating.

One form of energization at the magnetopause results from the acceleration by the electric field component tangent to the magnetopause current [e.g., *Reiff et al.,* 1977; *Cowley,* 1980; *Cowley,* 1982]. In the one-dimensional current sheet approximation, with nearly oppositely directed magnetic fields on either side of the current layer and a small magnetic field component normal to the layer, particles that cross the current sheet will gain a parallel speed equal to roughly twice the *ExB* drift speed in the layer. This gain in parallel speed is typically on the

order of a few hundred km s^{-1}, which is substantial (compared to typical thermal speeds) for magnetosheath and ionospheric ions, but is not significant for electrons. Acceleration by the electric field in the spacecraft frame is identical to the acceleration calculated by accounting for the particle motion in the deHoffman-Teller reference frame (for which the tangential electric field is zero) with suitable transformations between this reference frame and the spacecraft frame [e.g., *Cowley et al.,* 1982]. This current sheet acceleration contributes to the distinctive D-shaped ion distributions that have commonly been observed near the dayside magnetopause [e.g., *Smith and Rodgers,* 1991; *Fuselier et al.,* 1991]. Numerous examples have been shown where spacecraft data indicate good agreement with the predicted bulk acceleration [e.g., *Paschmann et al.,* 1990; *Sonnerup et al.,* 1990, 1995; *Phan et al.,* 1996; *Scudder et al.,* 1999]. This acceleration can at times be large enough to cause the plasma in the LLBL to flow in the opposite direction from the adjacent magnetosheath [*Gosling et al.,* 1990a]. These accelerated flows are observed in the immediate vicinity of the magnetopause and on the most recently reconnected field lines. Deeper in the magnetosphere, the LLBL flow is typically slower than in the magnetosheath and intermediate between the magnetosheath and magnetospheric flow speeds.

In addition to the current sheet acceleration that occurs at the magnetopause, some core or bulk heating is also observed [e.g., *Gosling et al.,* 1996; *Thomsen et al.,* 1997; *Paschmann et al.,* 1993]. This heating is observed in both the electrons and the ions, and tends to increase the temperature more in the direction parallel to the local magnetic field than in the perpendicular direction. Magnetosheath ions are typically highly anisotropic with $T_\perp > T_\parallel$, and the heating at the magnetopause reduces this anisotropy. Magnetosheath electrons are more nearly isotropic or slightly isotropic ($T_\perp > T_\parallel$), and the heating at the magnetopause creates the opposite sense of anisotropy ($T_\parallel > T_\perp$). This change in the electron temperature anisotropy was found to be the most reliable signature of the magnetopause location at low-shear magnetopause crossings, where the magnetic field measurements alone often do not give a clear indication of the magnetopause location [*Paschmann et al.,* 1993].

Some fraction of the incident plasma at the magnetopause is transmitted across the current sheet and some fraction is reflected. Both the transmitted and reflected portions of the particle distributions undergo the above-mentioned acceleration and heating in their encounter with the current sheet. Although the mechanism responsible for plasma reflection is not well known, observations suggest that the reflection coefficient can be substantial, on the order of about 50% [*Fuselier et al.*, 1991; *Onsager et al.*, 1993; *Lockwood et al.*, 1994].

Following acceleration at the current sheet, the reflected and transmitted components stream away from the magnetopause along the magnetic field. The LLBL is therefore characterized by the mixture of plasma consisting of the magnetospheric and ionospheric plasma inside the magnetosphere that have not encountered the magnetopause, plus the transmitted magnetosheath plasma and the reflected magnetospheric and ionospheric plasma that were accelerated at the magnetopause. Due to the disparate densities and temperatures of these different source populations, the resulting plasma distributions in the boundary layers can be clearly observed [e.g., *Paschmann et al.*, 1989; *Fuselier et al.*, 1991] and shown to match closely the predicted distribution functions [*Cowley*, 1982].

The properties of the LLBL away from the magnetopause are also influenced by the accessibility of plasma from the magnetopause along the recently reconnected field lines. As the plasma streams away from the reconnection site and from the magnetopause along the magnetic field, it also drifts perpendicular to the magnetic field in the direction of the bulk plasma convection. Therefore, just inside the separatrix on the most recently reconnected field lines, only those particles with the highest parallel speeds will be observed away from the magnetopause. The magnetosheath electrons have higher speeds than the ions, and therefore, the outer edge of the LLBL (just inside the separatrix on the reconnected field lines) will be populated by magnetosheath electrons but not magnetosheath ions. Furthermore, a low-speed cutoff will be observed in the entering magnetosheath populations, below which particles from the magnetosheath do not have access. As an observer moves from the separatrix toward the magnetopause in the direction of the ExB drift, this low-speed cutoff will move to lower speeds. The electron and ion low-speed edges on the distributions that are formed by this velocity-filter effect have been observed in the LLBL [e.g., *Gosling et al.*, 1990b] and in the MSBL at high latitudes due to northward IMF reconnection [e.g., *Chandler et al.*, 1999; *Onsager et al.*, 2001].

The large-scale properties of the boundary layers that form through reconnection are also affected by magnetospheric convection that is driven by the reconnection process. When the IMF is southward and reconnection occurs at low latitudes, convection of the newly reconnected magnetic flux transports the plasma sunward in the dayside magnetosphere toward the magnetopause and then to high latitudes over the poles and into the magnetotail. Any boundary layer that previously

existed inside the magnetosphere or is forming through reconnection is removed by the convection of the plasma toward the magnetopause and its subsequent downtail motion. Therefore while reconnection is forming the LLBL by heating and intermixing the plasma across the magnetopause, it is also simultaneously removing the LLBL through the convection of the reconnected flux.

On the other hand, when reconnection occurs at high latitudes under northward IMF, magnetosheath field lines become connected to Earth's field in one hemisphere, with the field line draping over the dayside magnetosphere and extending along the magnetopause to high latitudes in the other hemisphere. A second reconnection in the other hemisphere can then create a new closed field line that lies just outside the magnetopause [e.g., *Song and Russell*, 1992; *Le and Russell*, 1996]. As reconnection occurs during northward IMF, the boundary layer should continuously build up as new closed field lines are added to the dayside magnetopause [e.g., *Paschmann et al.*, 1990].

With a nonzero y-component of the IMF, the newly formed closed field lines will initially have some shear relative to the magnetospheric field. Therefore immediately following northward IMF reconnection, either in one hemisphere or in both hemispheres, the initial boundary layer at low latitudes will be located outside the magnetopause, and classified as the MSBL, rather than as the LLBL. As the information that reconnection has occurred propagates along the magnetic field via Alfven waves, the magnetic shear between the boundary layer field and the magnetospheric field will diminish, and the shear between the boundary layer field and the magnetosheath will increase. When the dominant rotation between the magnetosheath and the magnetospheric field becomes located outside (sunward of) the boundary layer, the boundary layer would then be classified as the LLBL [e.g., *Song and Russell*, 1992; *Fuselier et al.*, 1997]. Thus, the magnetopause would move through the boundary layer as the magnetic field shear gradually relaxed on the newly formed closed field lines. For a magnetosheath Alfven speed of about 200 km s^{-1}, this relaxation of the boundary layer field to a magnetospheric orientation would require about 7 min as the Alfven wave propagates roughly 15 Earth Radii (R$_E$) from the high-latitude reconnection site to the equator.

The role of reconnection in forming the LLBL during northward IMF has been much less clear than its role during southward IMF. Spacecraft that cross the magnetopause at low latitudes during northward IMF often do not observe the clear kinetic signatures of reconnection that are typical during southward IMF, and the relationship between the LLBL and particle energization at the magnetopause has not been resolved. The anisotropic particle heating mentioned above has been found to be a good indicator of the location of the magnetopause, and is particularly useful as an observational signature of the magnetopause and the LLBL under northward IMF conditions when the magnetic shear at the low-latitude magnetopause can be small. However, an important outstanding question regarding the northward IMF LLBL is whether the LLBL forms through local heating of magnetosheath plasma at the low-shear magnetopause or remote heating at the high-latitude, high-shear magnetopause.

In this paper, we present observations made by the Polar spacecraft at the high-latitude magnetopause that illustrate the role of high-latitude reconnection for forming the LLBL under northward IMF conditions. These observations show clear signatures of reconnection occurring at high latitudes. The plasma energization at the magnetopause and the velocity-space cutoffs in the electrons and ions both inside and outside the magnetopause on the recently reconnected field lines are clearly observed. A key aspect of the reconnection-formed boundary layers described here is the strong local heating measured *in situ* at the high-latitude magnetopause current layer. These observations demonstrate that heating of the magnetosheath plasma at the high-latitude magnetopause where the magnetic shear is high is sufficient to account for the LLBL properties observed at low latitudes where the local magnetic shear is low. Coupled with previous results showing that high-latitude reconnection can create new closed field lines on the dayside magnetosphere [e.g., *Le and Russell*, 1996; *Onsager et al.,* 2001], these observations demonstrate that reconnection at high latitudes can account for both the heating of the magnetosheath plasma that populates the LLBL and for the trapping of the heated plasma on the new closed field lines that become the LLBL.

2. OBSERVATIONS OF BOUNDARY LAYER FORMATION THROUGH HIGH-LATITUDE RECONNECTION

The Polar spacecraft has made a number of crossings of the high-latitude magnetopause while the IMF was steadily northward. These crossing have allowed the detailed investigation of high-latitude reconnection and boundary layer formation as the magnetopause repeatedly moved across Polar. This paper presents observations of the plasma at these magnetopause crossings that were made by the Hydra particle spectrometer onboard Polar.

The Hydra instrument measures the three-dimensional electron and ion velocity-space distribution functions using 12 narrow field-of-view detectors spread over the unit sphere [*Scudder et al.*, 1995]. The distribution functions described here cover the energy range from 5 eV/q to 20 keV/q with a temporal resolution of 13.8 s. These 13.8-s distributions are averages of the ion and electron energy sweeps made over the entire energy range in 1.15 s. It has been assumed in the analysis that all ions are protons. The magnetic field measurements were obtained from the Polar Magnetic Fields Investigation [*Russell et al*, 1995] that measures the vector magnetic field at 8-Hz resolution.

A 30-min interval of the plasma and magnetic field measurements during multiple crossings of the magnetopause on April 11, 1997 are shown in Plate 1. The IMF had been steadily northward with a B_z magnitude of around 15-20 nT (not shown) for roughly 8 hours prior to these observations. The panels in Plate 1 contain, from top to bottom, (a) the ion temperature; (b) the electron temperature; (c) the electron density; (d-f) the x, y, and z components of the magnetic field (GSM coordinates); (g) the parallel electron differential energy flux (0°-30° pitch angles); (h) the anti-parallel electron flux (150°-180° pitch angles); (i) the parallel ion flux; and (j) the anti-parallel ion flux. Two intervals when Polar was in the magnetosheath occurred from about 1435 to 1442 UT and 1446 to 1451 UT. The magnetosheath plasma can be identified by its low temperature, its high density, and by the northward IMF (Plate 1f). Crossings of the magnetopause current sheet are clearly seen by the strong rotations in B_z and B_x and smaller rotations in B_y. Near the current sheet crossings, the electrons and ions heat abruptly, and the density decreases with distance into the magnetosphere. At the times corresponding to the beginning and end of the data shown in Plate 1, the density was roughly two orders of magnitude lower than in the magnetosheath. In the boundary layers surrounding the magnetopause current sheet, the density was slightly reduced relative to the magnetosheath, but at the higher energies (a few hundred eV for electrons and a few keV for ions) the observed fluxes were well above the fluxes detected both in the magnetosheath and in the magnetosphere.

The fact that reconnection was actively taking place and forming the boundary layers is evident from the spatial layering seen in the parallel and antiparallel electron and ion fluxes. For example, the magnetopause current layer was crossed outbound at about 1434:15 UT, as seen from the sharp rotation in B_x and B_z. Electron and ion distribution functions showing contours of constant phase space density at four time intervals during this current sheet crossing are given in Figure 1. The lower two rows contain the electron and ion distributions as functions of parallel and perpendicular speed. The outermost electron contour corresponds to a phase space density of 10^{-30} s^3 cm$^-$

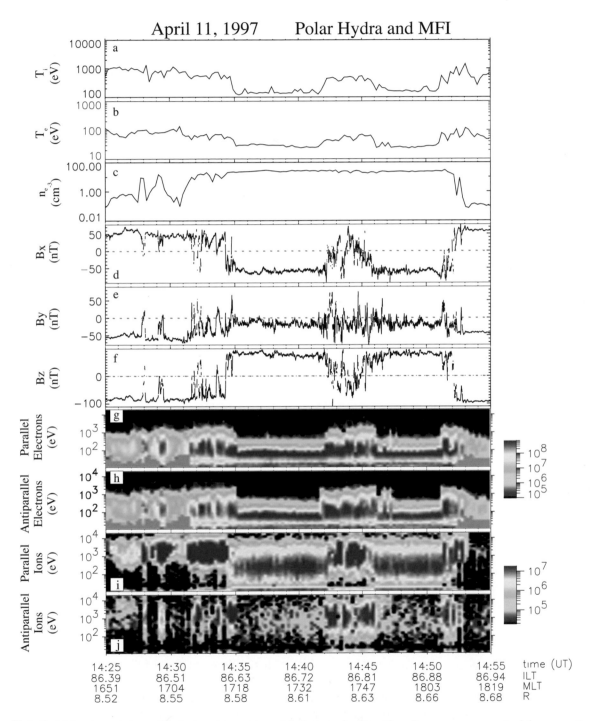

Plate 1. (a) Ion temperature; (b) electron temperature; (c) electron density; (d, e, f) *x, y, z* component of the magnetic field (GSM coordinates); (g) electron flux with 0°-30° pitch angles; (h) electron flux with 150°-180° pitch angles; (i) ion flux with 0°-30° pitch angles; and (j) ion flux with 150°-180° pitch angles.   Electron and ion flux is in units of differential energy flux (cm^{-2} s^{-1} sr^{-1} ΔE^{-1} E).   The times when Polar was in the magnetosheath are identifiable from the low temperature, high density, and the northward (+z) orientation of the magnetic field.

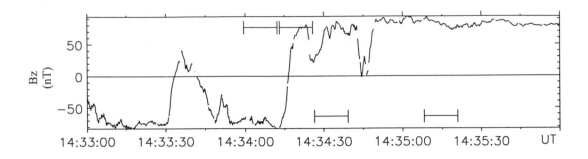

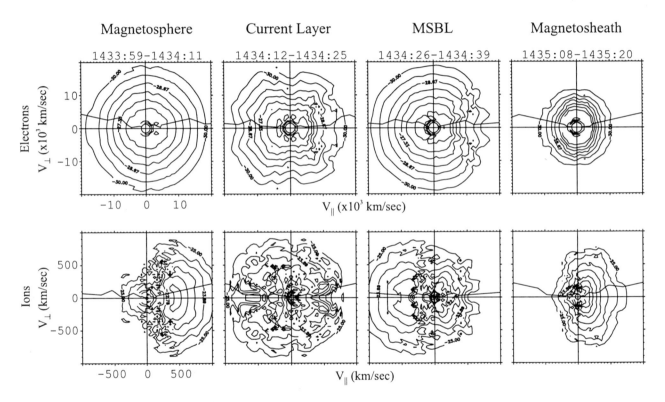

Figure 1. (top) Magnetic field component in the *z* GSM direction; and (bottom) electron and ion phase space density versus parallel and perpendicular speed during a crossing of the magnetopause current sheet. The two-dimensional distributions are plotted in a reference frame moving with the perpendicular bulk velocity. The perpendicular bulk ion velocity was, from left to right, 130, 27, 47, and 100 km/s. The dotted lines indicate the minimum pitch angle measured for each distribution function.

[6], and the outermost ion contour corresponds to 10^{-25} s^3 cm$^-$[6]. The dotted line near the $v_\parallel$ axis indicates the minimum pitch angles resolved by the Hydra instrument over the measurement interval. These observations are organized in energy – pitch angle space, with the symmetry between $v_\perp$ and $-v_\perp$ imposed in the plots. These distribution functions are plotted in a reference frame moving with the bulk convection speed perpendicular to the magnetic field. The *z* component of the magnetic field is shown in the top panel,

with the times of the 13.8-s distributions indicated with brackets.

Just inside the current layer in the magnetosphere (distributions from 1433:59 UT), the electrons were hot and isotropic, while the ions were highly anisotropic and consisted primarily of a hot D-shaped population flowing parallel to the magnetic field (toward the ionosphere from the magnetopause). Within the current layer, where the major rotation in the magnetic field occurred (1434:12

UT), the electron and ion distributions were hot, relatively isotropic, and showed considerable small-scale structure. Because Polar was in the current layer for only a brief time interval, it is possible that some time aliasing occurred over the 13.8-s measurement interval. Therefore although the small-scale structure may not be accurately resolved, the distributions show clear evidence of strong heating in the current sheet.

Just outside the current layer (1434:26 UT) the ion distribution consisted of two distinct populations. The ion population flowing in the parallel direction (toward the magnetopause) was relatively cool and dense, while the ion population in the antiparallel direction was relatively hot and closely resembled the parallel ions measured in the magnetosphere (1433:59 UT). The electron distribution at this time had a temperature that was similar to inside the current layer (1433:59 UT). Finally, upstream from the magnetopause in the magnetosheath (1435:08 UT), the electron and ion distributions had much lower temperatures and a $T_\perp > T_\parallel$ anisotropy.

The observations shown in Figure 1 indicate that strong heating of the magnetosheath electrons and ions occurred at the high-latitude magnetopause while reconnection was occurring poleward of the spacecraft. The heated magnetosheath plasma then streamed away from the magnetopause on the interconnected field lines. Outside the magnetopause in the MSBL, heated ions were observed in the antiparallel direction (into the magnetosheath), while inside the magnetosphere, heated ions were observed in the parallel direction (into the magnetosphere). The electrons in the magnetosphere were highly isotropic even though the source of heating was the magnetopause, since they have sufficient time with their high speeds to travel from the magnetopause to lower altitudes where they mirror and return to the magnetopause.

An interesting feature of the electrons observed in the MSBL (outside the magnetopause at 1434:26 UT) is the lack of electrons in the parallel direction with pitch angles less than about 45°. If reconnection had occurred only in the Northern Hemisphere above the spacecraft, the electrons heated at the magnetopause would only be expected to be observed flowing in the antiparallel direction (away from the magnetopause). On the other hand, if reconnection also occurred in the southern hemisphere forming new closed field lines, heated electrons should also be observed in the parallel direction (arriving from the Southern Hemisphere reconnection site) [e.g., *Onsager et al.,* 2001]. The MSBL electron distribution in Figure 1 does not fit well with either of these scenarios. One possible interpretation for the observed lack of low pitch-angle electrons is that reconnection had occurred in only the Northern Hemisphere, but the MSBL electrons had mirrored equatorward of the spacecraft and returned to the magnetopause. It is expected that the magnetic field is higher near the equatorial plane than at high latitudes, and therefore this mirroring at low latitudes could account for the loss-cone feature seen in the heated electrons. A cutoff at pitch angles around 45° would result from a magnetic field strength at the equator that is a factor of two higher than at the high-latitude magnetopause, which is consistent with estimates made from gas-dynamic and convected-field calculations (e.g., Figures 10 and 12 of *Spreiter and Stahara* [1985]). The low pitch-angle particles that do not mirror near the equatorial plane will be lost to interplanetary space as long as reconnection in the Southern Hemisphere has not yet occurred.

An inbound crossing of the magnetopause current layer is shown in Figure 2. Upstream from the MSBL in the magnetosheath (1441:20 UT), the electrons and ions were cold, with the typical $T_\perp > T_\parallel$ anisotropy of the ions and more nearly isotropic electrons. When the MSBL was encountered upstream from the magnetopause (1441:48 UT), heated electrons were observed streaming in the antiparallel direction, indicating that reconnection has occurred above the spacecraft. This measurement was made in the outer edge of the MSBL, where the heated ions from the magnetopause were not observed, presumably due to their lower speeds. Deeper in the MSBL, i.e., closer to the magnetopause (1442:02 UT), the electron distribution was more nearly isotropic, and the heated ions were observed streaming away from the magnetopause into the boundary layer. A lack of electrons at pitch angles less than about 45° was observed at this time, similar to the MSBL electron distribution shown in Figure 1. Note that there is also a highly field-aligned electron enhancement in this MSBL distribution at low pitch angles; however, those low pitch angles were not well resolved during this 13.8-s measurement interval (indicated with the dotted line). Finally, deeper within the current layer (1442:15 UT), both the electron and ion distributions were fully heated and nearly isotropic.

An important point to note in Plate 1 and in Figures 1 and 2 is that substantial heating is observed at the current sheet, and the heated electrons and ions stream both into the magnetosphere and back into the magnetosheath on the interconnected field lines. At low latitudes in the LLBL, this boundary layer plasma will be observed between the magnetosheath and the magnetospheric plasma. Although observed at low latitudes where the local magnetic shear is

Formation of the LLBL in the Context of a Unifying Magnetopause Reconnection Mechanism

Athanasios Boudouridis

Department of Atmospheric Sciences, University of California, Los Angeles

Harlan E. Spence

Center for Space Physics, Boston University, Boston

T. G. Onsager

Space Environment Center, NOAA, Boulder, Colorado

A recently proposed unifying magnetopause reconnection mechanism combines elements of the Bursty Single X-line and Multiple X-line Reconnection models. This reconnection scenario is based on ion precipitation data from two co-orbiting Defense Meteorological Satellite Program (DMSP) spacecraft (F6 and F8). The two energy dispersions observed can be explained by two separate injections resulting from two bursts of magnetopause reconnection. However, their observed overlap cannot be explained with the conventional model of bursty magnetopause reconnection, but requires additional assumptions which form the basis of the generalized reconnection scheme. In the framework of this unifying model, the observed electron features of a mixed magnetosheath and magnetospheric population can be explained in terms of the fossil Flux Transfer Event picture. In doing so this model provides a direct way for the formation of the Low-Latitude Boundary Layer (LLBL) via magnetic merging, on both open and closed field lines, at the dayside magnetopause during southward IMF conditions with a strong B_y component.

1. INTRODUCTION

A fundamental, ongoing debate on the nature of the high-altitude Low-Latitude Boundary Layer (LLBL) focuses on the topology of the magnetic field lines threading it. One view is that the LLBL exists on closed field lines. Various mechanisms have been proposed to introduce magnetosheath plasma onto these field lines, including diffusion across field lines, patchy reconnection [e.g., *Nishida*, 1989], and reconnection poleward of the cusp under northward Interplanetary Magnetic Field (IMF) conditions [e.g., *Song and Russell*, 1992]. For a review see *Lotko and Sonnerup* [1995]. The opposite view supports an open LLBL model in which the newly reconnected field lines evolve smoothly from

Earth's Low-Latitude Boundary Layer
Geophysical Monograph 133
Copyright 2003 by the American Geophysical Union
10.1029/133GM13

LLBL, just inside the magnetopause, to cusp to mantle configuration. At low altitudes the LLBL forms a continuous velocity dispersion ramp with the cusp ions, while its different nature arises from the free mixing of the magnetosheath and magnetospheric plasma immediately after reconnection (see extensive reviews by *Onsager and Lockwood* [1997] and *Lockwood* [1998]). Recently, *Newell and Meng* [1998] in an effort to reconcile the two opposing ideas proposed a mechanism that invokes a competition between diffusion and merging. This creates an open LLBL due to reconnection near noon, but a closed LLBL due to diffusion away from noon.

Boudouridis et al. [2001] developed a magnetopause reconnection model that unifies the the Bursty Single X-line Reconnection (BSXR) [*Scholer*, 1988; *Southwood et al.*, 1988] and Multiple X-line Reconnection (MXR) [*Fu and Lee*, 1985; *Lee and Fu*, 1985] models. This reconnection scenario operates under southward IMF conditions with a strong B_y component. The premise of the model, called Bursty Multiple X-line Reconnection (BMXR), is the overlapping ion energy-latitude dispersions observed by two co-orbital Defense Meteorological Satellite Program (DMSP) spacecraft. Consequently, *Boudouridis et al.* [2002] interpreted the electron features observed at the same time as signatures of "fossil" Flux Transfer Events (FTEs) and consistent with the BMXR mechanism.

The advantage of the BMXR mechanism is that it predicts the presence of a mixed magnetosheath and magnetospheric plasma population on both open and closed magnetic field lines. It therefore provides a convenient process for the formation of the LLBL on both magnetic topologies through a single unifying mechanism. In section 2 we present the DMSP observations. Section 3 describes the essentials of the BMXR mechanism derived on the basis of the ion data, while section 4 incorporates the electron data in this picture. Finally, section 5 discusses the formation of the LLBL on open and closed field lines in the context of the BMXR model.

2. DMSP OBSERVATIONS

The two spacecraft, F6 and F8, are in sun-synchronous, circular polar orbits and are virtually co-orbital in the dawn-dusk plane. They orbit at ~850 km, with F6 being slightly faster than F8, resulting in a race-track effect. Our study takes place during a southern hemisphere polar pass, around 1040 UT on January 10, 1990. The spacecraft move from high to low latitude, in the dawn side of the polar ionosphere. F6 is leading F8 with a separation time of about 60 seconds. They

both measure precipitating particle fluxes using identical SSJ/4 curved plate electrostatic analyzers, in 20 energy channels logarithmically spaced over the energy range of 30 eV to 30 keV. A complete 20 point electron and ion spectrum is obtained every second, corresponding to a spatial resolution of approximately 7 km along the orbital track. For a more detailed account of the spacecraft orbits and the instruments onboard see *Hardy et al.* [1984].

The three IMF components during our event, measured by the IMP 8 spacecraft, were $(B_x, B_y, B_z) =$ (-5 nT,+8 nT,-3 nT) and remained steady throughout the event. Our observations therefore occur during southward IMF conditions with a big B_y component, both favorable for the BMXR mechanism as discussed later on. Plate 1 shows integral and differential energy fluxes of ions (a) and electrons (b) from both spacecraft as a function of magnetic latitude. The line color of the integral fluxes refers to the signatures of the various magnetospheric regions encountered (shown immediately above them), as determined from the particle fluxes using the Newell-Meng criteria [*Newell and Meng*, 1988].

Here we concentrate on the ion data leaving the electron data for section 4. Two energy-dispersed features can be seen in the F6 data (top), indicated by the arrows 1 and 2, with equatorward edges at 76.4° and 77.2° magnetic latitude. Each extends poleward by several degrees. The same ion dispersions appear on the data from the F8 spacecraft (bottom pannels) which passed through the same region a minute later. This time, however, the equatorward edges of the dispersions have moved poleward to 76.7° and 77.5° magnetic latitude respectively. This implies a poleward motion of the energy dispersions of about 0.3° per minute or an equivalent latitudinal velocity of 600 m/s.

The interesting feature of these dispersions is their clear latitudinal overlap [*Boudouridis et al.*, 2001; 2002] observed between 77.2° (77.45°) and 77.7° (77.6°) in the F6 (F8) data. This overlap is not predicted by the conventional models of magnetopause reconnection [e.g., *Lockwood and Smith*, 1994; *Lockwood and Davis*, 1996]. The BSXR model, for example, depicts two different injections residing on explicitly distinct set of field lines that move away from the subsolar point one after the other. Unless the plasma on these field lines violates the frozen-in condition to move from one line to the other, it can never produce the overlapping dispersions seen at low-altitudes. A number of possible mechanisms to produce such an overlap have been proposed in recent years (for a review see *Boudouridis et al.* [2001]).

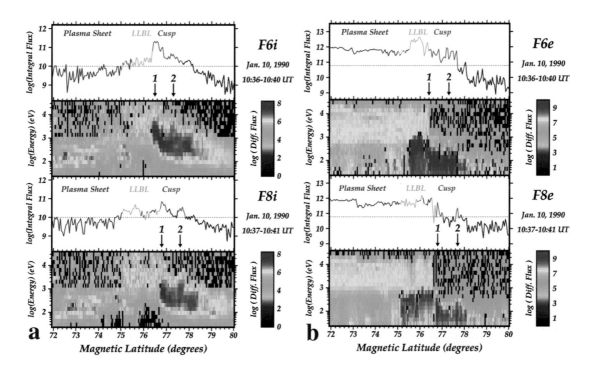

Plate 1. Integral and differential ion (a) and electron (b) energy fluxes. The line color of the integral fluxes corresponds to the regions shown immediately above. The black part corresponds to any other region not identified by the Newell-Meng criteria. The arrows point the location of the high-energy part of the energy dispersions.

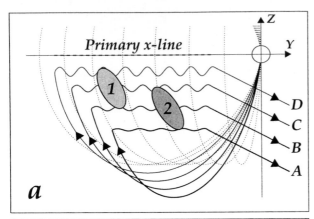

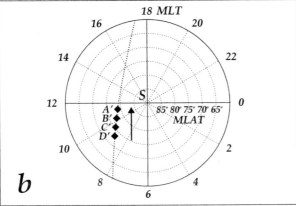

Figure 1. Reconnection snapshot illustrating the mechanism producing the overlap of the ion energy dispersions. (a) The plasma "blob" locations at the magnetopause, viewed from the sun with Earth at the upper right corner. (b) The respective feet in the polar ionosphere of the field lines bounding them. The arrow indicates the apparent motion of the energy features. The overlap occurs between the lines B and C, connecting to B′ and C′ respectively in the ionosphere [*Boudouridis et al.*, 2001].

3. THE BMXR MECHANISM

Boudouridis et al. [2001] argued that the above overlapping dispersions are easily generated through the BMXR mechanism, a simultaneous action of the BSXR and MXR processes with different temporal and spatial scales, together with different contributing reconnection rates. In this scenario MXR forms a semi-continuous, global, low level background of magnetopause reconnection. BSXR instead, is intermittent, patchy, and in the form of enhanced reconnection bursts.

The resulting picture is illustrated in Figure 1 taken from *Boudouridis et al.* [2001]. Panel (a) shows the view from the sun of multiply-reconnected magnetopause field lines as they move duskward (in the southern hemisphere for positive IMF B_y) and antisunward under the ambient magnetosheath flow and the tension

on the newly reconnected field lines. These form the MXR background which serves as a topological link for otherwise longitudinally distant regions of the dayside magnetopause. It is clear from Figure 1a that a significant IMF B_y component is a necessary factor for the creation of these structures. Its absence will lead to the formation of isolated magnetic islands with little or no longitudinal connection.

On top of this process, the presence of a localized and time-dependent enhanced resistivity at the central, primary neutral line will modulate the rate of reconnection there, switching to the BSXR mechanism in the way outlined by *Shi et al.* [1991] (see Figure 2). A layer of singly reconnected field lines will form, piling-up around the multiply reconnected ones as they move more or less in the same direction. Two longitudinally confined injections of this type acting at different longitudes, lead to the formation of plasma "blobs" which lie on the same multiply-reconnected field lines produced by the MXR process. Their plasma will escape along these lines down towards the ionosphere producing the two overlapping ion energy dispersions observed by the DMSP spacecraft. The longitudinal confinement of the "blobs" is necessary to produce two ion dispersions each with its own history, controlled by the reconnection burst at the magnetopause. Two extended injections will result in the merging of the two dispersions in one. Any mechanism that increases the resistivity at the magnetopause locally can be responsible for such a

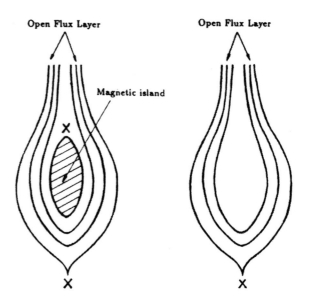

Figure 2. Field line configuration at the magnetopause looking from dusk due to the BSXR process (right) and the combined MXR and BSXR processes (left). The X denotes the x-line positions. Adapted from *Shi et al.* [1991].

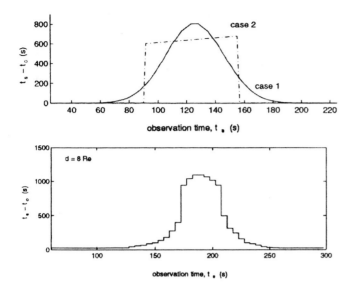

Figure 3. Time elapsed since reconnection $(t_s - t_o)$ as a function of observation time t_s. Top panel shows predictions for separate action of the BSXR (case 1) and MXR (case 2) models. Bottom panel shows the observed time during a satellite crossing of the reconnection bulge. Adapted from *Lockwood and Hapgood* [1998].

confinement. Panel (b) of Figure 1 shows the mapping of the multiply-reconnected field lines on the ionosphere and the region of dispersion overlap.

Corroborating evidence of the combined action of the MXR and BSXR processes comes from the modeling and observations reported by *Lockwood and Hapgood* [1998]. The top panel of Figure 3 gives predictions of the time since reconnection $(t_s - t_o)$, that would be deduced from measurements taken by a satellite moving in the north-south direction along the magnetopause (top to bottom in Figure 2) at observation time t_s, crossing a field line that reconnected at time t_o. Case 1 is for the BSXR model and case 2 for the MXR model. In case 1 the satellite encounters field lines with times $(t_s - t_o)$ that gradually increase from zero to a maximum value depending on how deep inside the bulge formed by reconnection the spacecraft goes, and similarly decrease to zero as it exits it. In case 2 the change in $(t_s - t_o)$ is abrupt as the spacecraft crosses into and out of the magnetic island more or less instantaneously. In the bottom panel of Figure 3 *Lockwood and Hapgood* [1998] presented the time since reconnection deduced during an actual crossing of the reconnection bulge. In the method used, they vary $(t_s - t_o)$ until the best fit of the observed density and temperature is obtained for every point along the satellite orbit. Due to the gradual variation of the time $(t_s - t_o)$ seen, they concluded that BSXR is the process at work at the magnetopause.

However, a closer look at Figure 3 reveals an additional feature. The observed time since reconnection may vary smoothly at first, but it jumps up by about 500 seconds at $t_s \simeq 170\,sec$ and then jumps down by about 400 seconds at $t_s \simeq 210\,sec$. This behaviour suggests a co-existence of the two reconnection models rather than one to the exclusion of the other. Looking back at the left drawing of Figure 2, the spacecraft first encounters the open flux layer. Time $(t_s - t_o)$ slowly increases as it moves from new to relatively older singly reconnected field lines. At some point it meets the magnetic island engulfed by them and $(t_s - t_o)$ has a discontinuous change. The reverse process occurs on its way out. *Lockwood and Hapgood* [1998] argued that "some kind of additional boundary process" would be needed for the MXR predicted time to vary smoothly at the edge of the island and "in the absence of a viable proposal for such a mechanism" they did not consider this model further. We argue that the combination of the two models provides this necessary "boundary-smoothing" mechanism.

4. ELECTRON DATA AND THE CLOSED LLBL

So far we were concerned with the ion energy dispersions. But what about the electron data? How do they fit in the BMXR mechanism? In Plate 1b we identify two electron injection features located poleward of ~76.3° (76.7°) and ~77.2° (77.6°) for F6 (F8), exhibiting less prominent energy-latitude dispersion than the ions due to the electrons' higher field-aligned velocity. The features have approximately the same equatorward edges and similar motion as the observed ion dispersions, as has been observed in past studies [*Newell and Meng*, 1988; *Onsager et al.*, 1993], and as required by cusp quasi-neutrality arguments [*Burch*, 1985]. All the properties of these features are consistent with the BMXR picture described above.

In addition to the two energy "dispersions", however, the electron data exhibit some other interesting features. These lie immediately equatorward of the two electron dispersions, between 75.5° (75.2°) and 76.2° (76.6°) for the F6 (F8) data. These magnetosheath-like features are identified as LLBL. They reside on the same field lines as plasma sheet electrons. This can be interpreted in two ways: (1) the field lines are open and the plasma sheet electron population is continuously replenished due to a number of mechanisms [*Onsager and Lockwood*, 1997], e.g., electron scattering at the magnetopause current sheet, magnetic mirroring due to larger field strength near the magnetopause crossing point rather than in the cusp, gradient and curvature

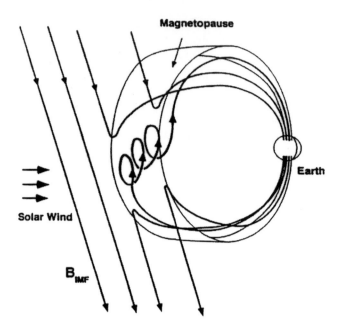

Figure 4. MXR process producing "fossil" FTE flux tubes on closed field lines [*Lee et al.*, 1993].

drifting of electrons from closed to open field lines, and electrostatic potential differences associated with maintaining quasi-neutrality; or (2) magnetosheath plasma has found its way onto closed field lines.

Onsager and Lockwood [1997] argued that in the first case "electron fluxes at low energies and large pitch angles would decay away over an extended region poleward of the initial loss of high energy, low pitch angle electrons." In Plate 1b we can see that the energy range of the plasma sheet electrons co-existing with magnetosheath electrons is the same as the energy range of the plasma sheet electrons further equatorward of the magnetosheath-like features, not lower. Moreover, these electrons are definitely low pitch angle electrons, since these are the only ones reaching the DMSP satellites. Finally, considering the high speed of the plasma sheet electrons and the fact that they are still present one minute later, we adopt the second view of magnetosheath plasma populating closed field lines.

Boudouridis et al. [2002] suggested that these features can also be understood in terms of the BMXR mechanism, together with the inclusion of the fossil FTE idea [*Lee et al.*, 1993]. The MXR process can result in FTE flux tubes with a mixture of magnetosheath and magnetospheric plasma, exclusively on either open or closed field lines [*Fu et al.*, 1990; *Lee et al.*, 1993]. In the open case, we expect the usual antisunward moving FTE structures. In the closed case, we expect FTEs moving slowly toward the Earth after reconnection has

possibly ceased. *Lee et al.* [1993] called these "fossil" FTEs and argued that they contribute to the formation and persistence of the closed LLBL during southward IMF conditions.

One such fossil FTE is shown in Figure 4, taken from *Lee et al.* [1993]. It is connected at both its ends to the terrestrial ionosphere, north in the dusk side and south in the dawn side. The tension force pulls the flux rope slightly toward dusk in the south and dawn in the north, as well as unwinding it. Since it went through multiple reconnections, it contains a significant amount of magnetosheath plasma together with the high energy plasma sheet component previously trapped on the closed field lines prior to reconnection. As was observed in the three-dimensional MXR simulations of *Fu et al.* [1990], this plasma will escape along the axis of the tube, and can account for the high-flux, dispersionless electron features equatorward of the main dispersions in Plate 1b.

5. DISCUSSION

Onsager and Lockwood [1997], briefly commenting on the existence of overlapping ion energy dispersions, conclude that "it is still unclear if the overlapping signatures are a complication of the open magnetosphere model or represent an entirely different mechanism." The BMXR mechanism provides a satisfactory solution to this problem by generalizing the open magnetosphere model to allow the generation of these low-altitude features. In addition, the mechanism presents a unifying approach within the open magnetosphere model for the formation of both the open and closed LLBL, without the complete exclusion of the closed LLBL [e.g., *Lyons et al.*, 1994; *Lockwood*, 1998] or the need for diffusion for its formation [e.g., *Newell and Meng*, 1998].

An important feature of the BMXR mechanism, and a consequence of the three-dimensionality of the underlying MXR process, is the breaking of the "degeneracy" of the Open-Closed Boundary, the boundary between the open and closed field lines. This now breaks into two different boundaries, separating three regions. The first region, lying equatorward of the dispersionless electron features, contains the closed field lines that were never opened by reconnection. The second region comprises the field lines that were opened and then closed again through the secondary x-lines. During their brief connection to the magnetosheath they are loaded with magnetosheath plasma, and even more plasma is deposited on them through the superimposed BSXR processes. On the other hand, if that connection is short enough ($\lesssim 10$ sec) they can still retain their plasma sheet component. This second region constitutes the low-altitude

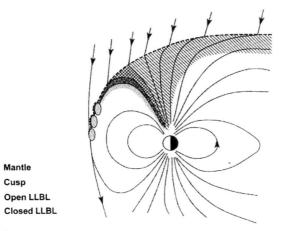

Mantle
Cusp
Open LLBL
Closed LLBL

Figure 5. Field and plasma structure at the vicinity of the cusp including the conclusions of the BMXR model. Adapted from *Crooker* [1977].

"image" of the high-altitude closed LLBL. Finally, the third region is formed by the open and poleward drifting magnetic field lines, replete with magnetosheath plasma which they deposit on the low-altitude ionosphere producing the energy-latitude dispersions.

It is clear from the two-point DMSP observations that the thickness of the intermediate region varies with time, oscillating between zero when the flux rope starts forming, and maximum just before the separation of the two magnetic field geometries [*Boudouridis et al.*, 2002]. This implies that the thickness of the respective region in the magnetosphere, the closed LLBL, is also variable. This is the favored portrayal of the high-altitude LLBL in previous studies [*Sckopke et al.*, 1981; *Mitchell et al.*, 1987]. *Mitchell et al.* [1987] reach a number of conclusions about the closed LLBL consistent with our observations. They note that for southward IMF and away from the subsolar point the LLBL size becomes more variable; both open and closed LLBL regions are present, but for intermittent merging the existence of the open kind is also intermittent.

We wish to close this work with an updated version of the dayside plasma structure presented by *Crooker* [1977] (see her Figure 3) shown in Figure 5. In this figure, which includes the ideas of the BMXR model, the closed LLBL consists of several flux ropes seen in cross section looking from dusk, whereas the open LLBL is formed by the singly-reconnected field lines surrounding them. It should be noted that such two-dimensional construction is not entirely appropriate for the inherently three-dimensional field structure of the BMXR model. The closed-LLBL flux ropes map to low altitudes at different longitude and not in the meridional plane as is schematically depicted in Figure 5.

As mentioned earlier, they connect the northern duskside ionosphere with the southern dawnside one, or vice versa depending on the sign of the IMF B_y component. A test of this prediction would be simultaneous north/south low-altitude particle observations on opposite sides (dawn/dusk) of the high-latitude ionosphere, that could detect signatures of the BMXR mechanism.

Finally, a thorough examination of low-altitude satellite data sets is required to test these ideas and establish the significance of the BMXR process to magnetospheric flux transfer. Similar overlapping dispersions have been seen in the past [*Boudouridis et al.*, 2001]. An excellent test of the model, however, would be observations of such structures with satellites at different altitudes and different time delays.

Acknowledgments. The authors wish to thank N. U. Crooker for her valuable comments. This work was supported by NASA grant NAG5-4273 and NSF grant ATM-9458424.

REFERENCES

Boudouridis, A., H. E. Spence, and T. G. Onsager, Investigation of magnetopause reconnection models using two co-located, low-altitude satellites: A unifying reconnection geometry, *J. Geophys. Res., 106,* 29,451, 2001.

Boudouridis, A., H. E. Spence, and T. G. Onsager, A new look at the pulsed reconnection model of the dayside magnetopause, *Adv. Space Res.,* in press, 2002.

Burch, J. L., Quasi-neutrality in the polar cusp, *Geophys. Res. Lett., 12,* 469, 1985.

Crooker, N. U., The magnetospheric boundary layers: A geometrically explicit model, *J. Geophys. Res., 82,* 3629, 1977.

Fu, Z. F., and L. C. Lee, Simulation of multiple X-line reconnection at the dayside magnetopause, *Geophys. Res. Lett., 12,* 291, 1985.

Fu, Z. F., L. C. Lee, and Y. Shi, A three-dimensional MHD simulation of the multiple X line reconnection process, in *Physics of Magnetic Flux Ropes, Geophys. Monogr. Ser., vol. 58,* edited by C. T. Russell, E. R. Priest, and L. C. Lee, p. 515, AGU, Washington, D. C., 1990.

Hardy, D. A., L. K. Schmitt, M. S. Gussenhoven, F. J. Marshall, H. C. Yeh, T. L. Schumaker, A. Huber, and J. Pantazis, Precipitating electron and ion detectors (SSJ/4) for the block 5D/flight 6-10 DMSP satellites: Calibration and data presentation, *Rep. AFGL-TR-84-0314,* Air Force Geophys. Lab., Hanscom Air Force Base, Mass., 1984.

Lee, L. C., and Z. F. Fu, A theory of magnetic flux transfer at the Earth's magnetopause, *Geophys. Res. Lett., 12,* 105, 1985.

Lee, L. C., Z. W. Ma, Z. F. Fu, and A. Otto, Topology of magnetic flux ropes and formation of fossil flux transfer events and boundary layer plasmas, *J. Geophys. Res., 98,* 3943, 1993.

Lockwood, M., Identifying the open-closed field line boundary, in *Polar Cap Boundary Phenomena,* edited by J. Moen, A. Egeland, and M. Lockwood, p. 73, Kluwer Academic Publishers, Dordrecht, Netherlands, 1998.

Lockwood, M., and C. J. Davis, On the longitudinal extent of magnetopause reconnection pulses, *Ann. Geophys., 14,* 865, 1996.

Lockwood, M., and M. A. Hapgood, On the cause of a magnetospheric flux transfer event, *J. Geophys. Res., 103,* 26,453, 1998.

Lockwood, M., and M. F. Smith, Low and middle altitude cusp particle signatures for general magnetopause reconnection rate variations, 1, Theory, *J. Geophys. Res., 99,* 8531, 1994.

Lotko, W., and B. U. Ö. Sonnerup, The low-latitude boundary layer on closed field lines, in *Physics of the Magnetopause, Geophys. Monogr. Ser., vol. 90,* edited by P. Song, B. U. Ö. Sonnerup, and M. F. Thomsen, p. 371, AGU, Washington, D. C., 1995.

Lyons, L. R., M. Schulz, D. C. Pridmore-Brown, and J. L. Roeder, Low-latitude boundary layer near noon: An open field line model, *J. Geophys. Res., 99,* 17,367, 1994.

Mitchell, D. G., F. Kutchko, D. J. Williams, T. E. Eastman, L. A. Frank, and C. T. Russell, An extended study of the low-latitude boundary layer on the dawn and dusk flanks of the magnetosphere, *J. Geophys. Res., 92,* 7394, 1987.

Newell, P. T., and C.-I. Meng, The cusp and the cleft/boundary layer: Low altitude identification and statistical local time variation, *J. Geophys. Res., 93,* 14,549, 1988.

Newell, P. T., and C.-I. Meng, Open and closed low latitude boundary layer, in *Polar Cap Boundary Phenomena,* edited by J. Moen, A. Egeland, and M. Lockwood, p. 91, Kluwer Academic Publishers, Dordrecht, Netherlands, 1998.

Nishida, A., Can random reconnection at the magnetopause produce the low latitude boundary layer?, *Geophys. Res. Lett., 16,* 227, 1989.

Onsager, T. G., and M. Lockwood, High-latitude particle precipitation and its relationship to magnetospheric source regions, *Space Sci. Rev., 80,* 77, 1997.

Onsager, T. G., C. A. Kletzing, J. B. Austin, and H. MacKiernan, Model of magnetosheath plasma in the magnetosphere: Cusp and mantle particles at low altitudes, *Geophys. Res. Lett., 20,* 479, 1993.

Scholer, M., Magnetic flux transfer at the magnetopause based on single X-line bursty reconnection, *Geophys. Res. Lett., 15,* 291, 1988.

Sckopke N., G. Paschmann, G. Haerendel, B. U. Ö. Sonnerup, S. J. Bame, T. G. Forbes, E. W. Hones, Jr., and C. T. Russell, Structure of the low-latitude boundary layer, *J. Geophys. Res., 86,* 2099, 1981.

Shi, Y., C. C. Wu, and L. C. Lee, Magnetic field reconnection patterns at the dayside magnetopause: An MHD simulation study, *J. Geophys. Res., 96,* 17,627, 1991.

Song, P., and C. T. Russell, Model of the formation of the low-latitude boundary layer for strongly northward interplanetary magnetic field, *J. Geophys. Res., 97,* 1411, 1992.

Southwood, D. J., C. J. Farrugia, and M. A. Saunders, What are flux transfer events?, *Planet. Space Sci., 36,* 503, 1988.

A. Boudouridis, Department of Atmospheric Sciences, University of California, Los Angeles, 405 Hilgard Avenue, Los Angeles, CA 90095. (e-mail: thanasis@atmos.ucla.edu)

H. E. Spence, Center for Space Physics, 725 Commonwealth Avenue, Boston University, Boston, MA 02215. (e-mail: spence@bu.edu)

T. G. Onsager, Space Environment Center, National Oceanic and Atmospheric Administration, 325 Broadway, Boulder, CO 80303. (e-mail: tonsager@sec.noaa.gov)

Antiparallel Reconnection as a Possible Source of High- and Low-Latitude Boundary Layers

A. Fedorov[1], E. Budnik[1], H. Stenuit, T. Moreau, and J.-A. Sauvaud

Centre d'Etude Spatiale des Rayonnements, Toulouse, France

Numerous remote and in situ observations indicate that magnetic reconnection is the dominant mode of solar wind - magnetosphere interaction on the dayside magnetopause, and probably the only mechanism responsible for the formation of both the high-latitude boundary layer (HLBL) and the low-latitude boundary layer (LLBL). To test this hypothesis, we have systematically ordered HLBL and LLBL observations from the Interball-1 spacecraft. To simplify the problem only data with the interplanetary magnetic field (IMF) close to the ecliptic plane have been used. To perform the study we constructed a global model of magnetopause topology based on the antiparallel merging hypothesis. This model predicts for arbitrary interplanetary conditions the location of the reconnection site on the magnetopause surface and a position of zones of rotational discontinuity (RD). All our LLBL and HLBL observations were classified into four types according to the observed plasma regimes. Then the location of each crossing was organized in a coordinate system defined by the simulated reconnection site and the zones of RD. We found that all types of HLBL and LLBL are well ordered in such a coordinate system. This study indicates that antiparallel field-line merging may be the main process responsible for the formation of all the observed types of boundary layer when IMF is about horizontal.

1. INTRODUCTION

It is generally believed that low-energy plasma observed in the boundary layer (BL) adjacent to the magnetopause is of the magnetosheath origin. Several entry mechanisms have been proposed to describe the transport of magnetosheath particles into the magnetosphere. (i) Magnetic merging/reconnection [*Dungey*, 1961; *Cowley and Owen*, 1989]. This process occurs locally, but has global consequences: once the interplanetary (IMF) and geomag-

netic fields become connected, they remain so while being traveled tailward along the magnetopause. Thus the magnetopause topology is changed over an elongated region, and magnetosheath plasma can continuously enter the magnetosphere over almost the entire path of reconnected field line. (ii) Impulsive penetration of magnetosheath plasma filaments on magnetospheric field lines [*Lemaire and Roth*, 1978; *Heikkila*, 1982]. (iii) Cross-field diffusion of magnetosheath plasma into the BL [*Eastman and Hones*, 1979]. Large-scale Kelvin-Helmholtz instability of the magnetopause surface has also been suggested as an alternative to reconnection to describe cross-magnetopause transport [*Miura*, 1987; *Fairfield et al.*, 2000].

The BL may be formed by a single mechanism, or by several different processes occurring under different conditions, or even simultaneously. Numerous evidences of plasma entry over the open field lines in the low-latitude boundary

[1] on leave from IKI RAN, Moscow, Russia

Earth's Low-Latitude Boundary Layer
Geophysical Monograph 133
Copyright 2003 by the American Geophysical Union
10.1029/133GM14

layer (LLBL) [*Gosling et al.*, 1990; *Fuselier et al.*, 1995] and in the high-latitude boundary layer (HLBL) [*Fedorov et al.*, 2001] point out the merging/reconnection as a preferable candidate for a single process. Here we use the term "reconnection" to indicate some (possibly unknown) physical mechanism which causes magnetically unconnected magnetospheric and magnetosheath regions become linked. Concerning the geometry and localization of the reconnection at the dayside magnetopause, there are two extremes: (i) The component model [*Gonzalez and Mozer*, 1974; *Sonnerup*, 1974], which suggests merging at the subsolar point where the solar wind dynamic pressure is high and the velocity is low. (ii) The antiparallel merging model [*Crooker*, 1979] predicts that reconnection occurs in the region where geomagnetic field and draped magnetosheath magnetic field adjacent to the magnetopause are oppositely directed and thus the magnetopause current is maximal. *In situ* observations of *Gosling et al.* [1996] and *Kessel et al.* [1996] of high-latitude reconnection during strongly nothward IMF are good evidences in favor of the last hypothesis.

In present study we test antiparallel merging as a process responsible for both HLBL and LLBL formation. For this we analysed INTERBALL-1 satellite magnetopause crossings in a coordinate system defined by the predicted position of instantaneous reconnection site. The coordinate system was constructed on the basis of a 3-D model of magnetopause topology. Since the difference between two reconnection concepts is most obvious when the IMF B_Y component is dominant, we have considered the magnetopause crossings occurring when the IMF $B_Y - B_Z$ vector was within $\pm 45°$ of magnetic equator.

2. A 3-D MODEL OF MAGNETOPAUSE TOPOLOGY

The location of the reconnection site of the antiparallel merging model depends strongly on the IMF direction. In order to analyse data in the light of this model, we need a reference frame which moves over the magnetopause surface in accordance with the position of the reconnection site. We note that, in addition to determining the location of the reconnection site, the convection of reconnected field lines over the magnetopause surface is also of key importance for boundary layer properties [*Cowley and Owen*, 1989]. For instance, positive B_Y IMF causes dawnward convection in the northern hemisphere HLBL, whilst with negative B_Y IMF the convection becomes duskward [*Gosling et al.*, 1985]. Thus the new coordinate system should also depend upon the reconnected field line convection pattern.

To define our frame of reference we constructed a model of the magnetopause topology which predicts the location of an antiparallel reconnection zone and the location of regions of rotational discontinuity (RD) across the magnetopause created by convection of reconnected field lines.

There are four main components of the model: the magnetopause shape, the magnetosheath plasma velocity and density, the magnetosheath magnetic field, and the magnetospheric magnetic field. We use the *Shue et al.* [1997] model of the magnetopause shape. The true spacecraft location was projected onto the model magnetopause along the radius-vector. Magnetosheath plasma flow parameters and magnetic field vectors were derived at the model magnetopause surface: the plasma parameters were derived from *Spreiter and Stahara* [1985], using data from the WIND plasma experiment [*Ogilvie et al.*, 1995]. The draped magnetosheath magnetic field was calculated using the technique proposed by *Alksne and Webster* [1970] together with data from the WIND magnetometer [*Lepping et al.*, 1995]. The magnetosheath plasma speed and number density, and the field magnitudes were then scaled to the values actually measured by Interball-1 in the magnetosheath adjacent to the magnetopause crossing. Magnetospheric magnetic field was calculated from the Tsyganenko 96 model [*Tsyganenko*, 1995]. Thus for any interplanetary conditions the model computes the distribution of magnetosheath velocity, number density, magnetic field vectors, and the distribution of magnetospheric field vectors over the entire frontside magnetopause. Then we determine the position on the magnetopause where there is maximum probability of antiparallel reconnection occurring. We define this as the zone where the angle between magnetosheath and magnetospheric magnetic fields is greater than $160°$. We restrict ourselves to reconnection occurring sunward of the terminator, although this assumption is not essential for the results of the study.

The model also determines the geometry of the RD regions created by reconnection. This merits some explanation. Figure 1 illustrates the evolution of a newly reconnected field line in case of high-latitude flank reconnection when the convection pattern is essentially 3-D. The magnetosheath plasma bulk velocity is high in this case, and the flow is perpendicular to the magnetic field. This figure displays a portion of the magnetopause surface with adjacent magnetospheric and magnetosheath flux tubes before and after reconnection. After reconnection the magnetosheath part of the newly reconnected flux tubes continue to move tailward with the magnetosheath flow velocity $\mathbf{V_S}$. At the same time the two "kinks" move at the Alfvén velocity V_A (in the magnetosheath flow frame) parallel (lines $\mathbf{S}$) and antiparallel (lines $\mathbf{N}$) to the magnetosheath magnetic field. In the unmoved frame the "kinks" travel with the velocity $\mathbf{V_S} \pm V_A \cdot \mathbf{b}$ along their "Alfvén wing" trajectories. Here $\mathbf{b}$ is the unit vector parallel to the magnetosheath magnetic field. The model computes all possible "kink" trajectories emanating from the reconnection region. It is worth noting that magnetosheath plasma entering the magnetosphere along magnetic $\mathbf{N}$ field lines forms the HLBL, and plasma entered along $\mathbf{S}$ lines creates the LLBL.

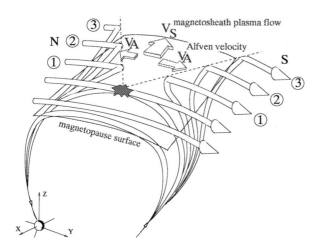

Figure 1. 3-D evolution of a field lines opened by high-latitude flank reconnection. The kink points of two newly opened field lines diverge from the diffusion region along **N** and **S** trajectories on the magnetopause surface. See text for details.

Figure 2 shows examples of the frontside magnetopause geometry calculated for interplanetary conditions during two encounters of Interball-1 with the magnetopause, on April 24, 1996 and February 15, 1996. On April 24 the IMF was dawnward and southward and the reconnection region **R** was predicted to be located at mid-latitudes on the dawn flank. On February 15 the IMF was purely duskward and the recon-

nection site **R** was located on the high-latitude dusk flank. If we assume that reconnection can occur in every point of the region **R**, then the area of the northern magnetopause surface containing all possible N-type and S-type kink trajectories is illustrated by the gray bands **N** and **S** in Figure 2. For simplicity we consider only the northern hemisphere; for these two particular directions of the IMF, regions of reconnection and of RD on the magnetopause also occur in the southern hemisphere.

Thus the regions **N** and **S** indicate the portions of the northern magnetopause with RD properties, with the magnetosheath field lines being topologically connected to respectively the northern cusp and to the southern cusp. Note that the bands show the regions where the magnetopause could be a RD; in reality a significantly smaller portion of the magnetopause may be open. Nevertheless the locations of the region **R** and of the bands **N** and **S** provide a basis for the construction of two new coordinate systems, to study the high-latitude and the low-latitude magnetopause. These new coordinate systems are introduced in Sections 3 and 4.

3. OVERVEW OF OBSERVATIONS AND STATISTICAL ANALYSIS OF THE HLBL

3.1. Data base

Interball-1 had an elliptical orbit with an apogee of 200000 km and period of 4 days. The present study is based on data obtained from the CORALL ion spectrom-

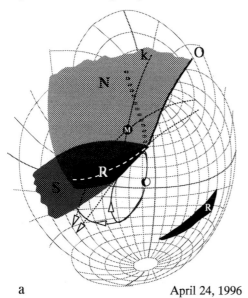

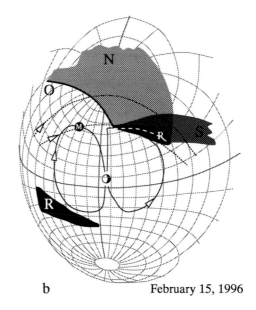

a   April 24, 1996   b   February 15, 1996

Figure 2. Two examples of the magnetopause topology computed for different interplanetary conditions. Each panel shows the frontside magnetopause surface with an embedded latitude-longitude GSM grid. The black spots **R** mark the reconnection regions of high magnetic shear. The grey bands **N** and **S** display possibly open parts of magnetopause in the northern hemisphere ; **N** is magnetically connected to the northern cusp and **S** to the southern cusp. Dotted black curves represent draped magnetosheath field lines. The line **k** shows the particular kink-point trajectory which passes through the spacecraft located at **M**. The dashed white curves in the regions **R** mark the median of the reconnection region. The line **O** indicates the kink point trajectory which originates from the point of the region **R** closest to the noon meridian.

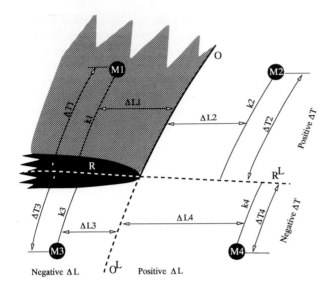

Figure 3. An schematic representation of the $\Delta L - \Delta T$ coordinate system. Labels are the same as for Figure 2. See text for details.

eter [*Yermolaev et al.*, 1997], the electron spectrometer Electron [*Sauvaud et al.*, 1997] and magnetometer MIF [*Klimov et al.*, 1995]. The IMF data [*Lepping et al.*, 1995] and solar wind data [*Ogilvie et al.*, 1995] monitored by the WIND spacecraft were obtained via CDAWeb ($http : //cdaweb.gsfc.nasa.gov$).

Seventy six examples of low-latitude and high-latitude magnetopause crossings were selected for analysis. The selection criteria were as follows: (i) The IMF clock-angle was within $\pm 45°$, that is, the magnitude of B_Y was greater than or equal to the magnitude of B_Z. (ii) Solar wind conditions were stable enough to exclude errors due to transient effects.

3.2. HLBL coordinate system

The location of the region **R** and the band **N** are used to construct the superficial coordinate system to study the high-latitude magnetopause. The reference line of a new system is boundary **O** (Figure 2). This boundary is the field line kink trajectory originated from reconnection occurring at the point of **R** closest to the noon meridian. It lies on the magnetopause, and separates the zones of rotational and tangential discontinuity. Note that the zone **N** lies dawnward of the boundary **O** on April 24 and duskward on February 15. So the satellite crossed the magnetopause (point **M**) inside the region **N** in the first case and outside it in the second.

The ΔL coordinate describes the location of the satellite with respect to the boundary **O**, as shown in Figure 3. Let $\mathbf{R}^L$ be the median line of the region **R**, and let **k1** be the the kink trajectory passing through the spacecraft position **M1**. For **M1** lying on the same side of the boundary **O** as **N** (and possibly inside **N**), ΔL is negative, with magnitude equal to the angular distance between the origin of the boundary **O**

and the intersection of **k1** with the median $\mathbf{R}^L$, that is, the distance along the white dashed line in Figures 2 and 3.

If the spacecraft position lies outside the **N** region (point **M2** in Figure 3), then ΔL is positive. In this case the "virtual" trajectory **k2** can be traced from **M2** back to the latitude corresponding to the origin of the line **O** (black dashed line R^L in Figure 3). If the satellite is located equatorward from the median of the region **R** (point **M3**) or equatorward from the line R^L (point **M4**) then the ΔL coordinate is calculated as angular distance between "virtual" kink trajectories traced from points **M3** or **M4** and "virtual" kink trajectories O^L traced back from the origin of the **O** boundary.

The second coordinate, ΔT, is defined as the time (in minutes) required for a kink point to propagate from the median line $\mathbf{R}^L$ to the spacecraft position. For the point **M1**, which is poleward of the median line $\mathbf{R}^L$ (Figure 3), ΔT is positive. For the point **M2** ΔT is the propagation time of a virtual kink point from R^L to **M2**. For points **M3** and **M4** ΔT are similarly defined, but are negative.

3.3. Statistical study of the HLBL crossings

To examine the 41 selected Interball-1 magnetopause crossings in the new coordinate system, we sorted them into 4 different classes according to the characteristics of the observed boundary layer (Plate 1). Thus: (i) Wide, stable, intense boundary layers (Plate 1a) up to $2 - 4\, R_E$ thick. This boundary layer is observed practically without gaps (no more then 30% of time) and demonstarates smooth transition to the weak low energy mantle deeper into magnetosphere. The ion properties show that the plasma observed in the layer adjacent magnetopause entered the magnetosphere locally along open field lines [*Fedorov et al.*, 2000]. (ii) Sporadic boundary layer (Plate 1b) with a set of ion and electron bursts. The gaps between particular bursts occupy more then 30% of time interval of HLBL observation. HLBL can be attached (as in example) or detached from the magnetopause. (iii) Thin boundary layer attached to the magnetopause (Plate 1c). The selection criteria for this case were: HLBL no longer then 10min of observation time; no gradual transition to the low energy mantle regime. (iv) No apparent HLBL (Plate 1d. Figure 4 displays the distribution of the selected cases on the frontside magnetopause. To indicate its class, each observation of the HLBL is coded in accordance with Plate 1). The distribution of the different types of HLBL appears to be random.

The same set of crossings plotted in $\Delta L - \Delta T$ coordinates shows a systematic pattern (Figure 5). This scatterplot was obtained in a following way: (i) For each case the geometry of the magnetopause topology for the existing interplanetary and magnetospheric conditions was computed. (ii) Then the $\Delta L - \Delta T$ coordinates of the point of the Interball-1 magnetopause crossing were derived.

Four important features appear immediately from Figure 5: (i) The stable wide BL with the local magnetosheath

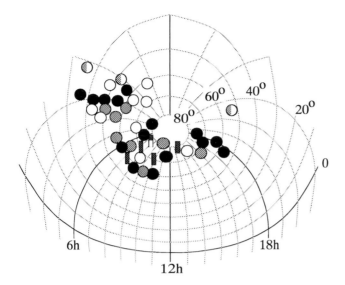

Figure 4. A front view of northern high-latitude magnetopause with an embedded latitude-longitude SM grid. The locations of Interball-1 magnetopause crossings are coded to indicate their observed HLBL plasma classification (see Plate 1). Half-filled circles correspond to intermediate cases between BL absent and sporadic BL.

plasma entry is observed essentially within the $-2h - -0h$ ΔL band (emphasized by light grey). (ii) The sporadic boundary layer is encountered at more negative ΔL values. (iii)The absence of the HLBL is also met in the $\Delta L < -2h$ region. But it extends further to more negative ΔL and to higher ΔT values. (iv) The thin HLBL is concentrated $\pm 1 Re$ around boundary **O**. (v) The stable BL is not observed in the reconnection region.

The model predicts the absence of BL for $\Delta L > 0$. Although the statistics are poor in this region, we do indeed see three cases of practically empty magnetosphere (lobes). The observation of a steady open magnetopause (marked by arrow) occurred when there was significant IMF B_X that possibly caused the disturbance of the magnetosheath magnetic field caused by the quasi-parallel bow shock occupying the subsolar region.

4. STATISTICAL ANALYSIS OF THE LLBL

4.1. Coordinate system for LLBL

Figure 6 shows two frontside magnetopause geometrical configurations with the predicted location of the reconnection sites (regions **R**) for the interplanetary conditions which occurred on a) Feb. 18, 1997 at 05:07 UT, and b) Feb. 13, 1998 at 19:38 UT. The point **M** is the location of the Interball-1 at the time of the magnetopause crossing. The curve **A** corresponds to the magnetospheric field line which touches the reconnection site closest to the solar meridian, and the curve

B shows the magnetospheric field line passing through the satellite. Figure 6 shows only the wing of RD created by open field lines passing over the low-latitude magnetopause (the Grey belt **S**). Note that in the two cases, (a) and (b), the spacecraft locations are similar, but the regions **R** are situated at significantly different latitudes due to differing IMF (in the X-Y plane for (a), and (0, -3.1, 3.1)nT GSM for (b)). In both cases the dusk reconnection is predicted to occur in the southern hemisphere because of the negative IMF B_Y.

Alfvén wings **S** shown in Figure 6 are associated with RD on open field lines linked to the north pole and convected tailward. Since in case (a), Feb. 18 1997, the field line **B** passes through the RD wing, we may expect this line to be open. In case (b), Feb. 13 1998, the first open field line **A** is located tailward of the spacecraft and the field line **B** has no topological connection to the **R** zone.

To perform a statistical analysis of LLBL in the reconnection frame, we introduce a new coordinate system, $\Delta\Lambda - \Delta\Theta$. $\Delta\Lambda$ is the angular separation of the points at which the field lines **B** and **A** intersect the GSM equator. The positive values of $\Delta\Lambda$ correspond to field lines which pass via either the reconnection zone **R** or the RD region **S**; negative values correspond to field lines which pass noonward of the field line **A**. $\Delta\Theta$ is the latitudinal distance between the satellite

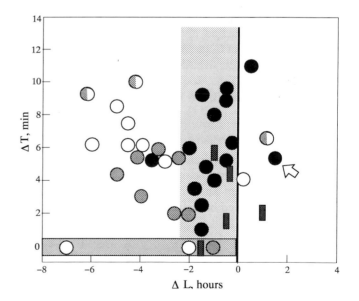

Figure 5. All high-latitude magnetopause crossings used in this study, presented in the $\Delta L - \Delta T$ coordinate system (see section 2 for explanation). The reconnection region is shown by horizontal grey band. The boundary **O** is marked by the thick vertical line $\Delta L = 0$. The vertical light grey rectanguler indicates the region where the stable open magnetopause was observed. Half-filled circles indicate cases for which there was almost no evidence of the BL, except for very rare plasma bursts. The observation marked by the arrow was for large IMF B_X.

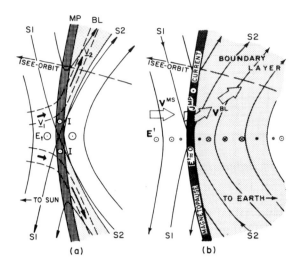

Figure 1. The topology of the magnetic field is identical for both cases; S1 and S2 represent different sheets of the separatix surface. (a) The reconnection geometry of the electric field is spatially constant, thus with zero curl. With a constant electric field and a reversing magnetic field direction (south to north) the plasma flows into the reconnection region from both sides; the outflow direction is to open field lines as shown. (b) With a plasma transfer event, since both fields **E** and **B** reverse (approximately) for a high-shear magnetopause, their cross product **E** × **B** is unidirectional. Most of the plasma can cross the moving magnetopause as a fluid in a dynamic process into the low latitude boundary layer, even onto closed magnetic field lines. Some plasma is lost from the system along the field lines.

Unfortunately, conditions for reconnection to occur are unknown [*Semenov et al.*, 1992, p4252; *Lockwood and Davis*, 1996, p865]. The reconnection electric field is always prescribed as an input parameter, often by a resistive MHD code [e.g. *Siscoe et al.*, 2001] that depends on an "anomalous" resistivity [*Coroniti*, 1985]. Although these theories concluded that "dayside reconnection is the dominant contributor ··· to the magnetospheric voltage" [*Cowley*, 1982], nevertheless there are many problems that cannot be explained by reconnection. As noted by *Phan et al.* [1997, p19,894]

> "Other processes [besides reconnection] must
> be responsible for transporting plasma across
> the local MP, onto the closed field LLBL.
> [These] are rather general occurrences, irre-
> spective of the IMF and the magnetic shear
> across the MP."

Similar concerns have been voiced by *Heikkila* [1978a]; *Lemaire et al.* [1979]; *Lundin and Evans* [1985]; *Newell and Meng* [1997], and others.

Axford and Hines [1961] offered an entirely different concept, that of a viscous-like interaction between the external solar wind and internal circulation of magnetospheric plasma. The term "viscous-like" was introduced as an analogy with fluid theory. It is by no means clear what meaning that term has in the magnetospheric context; all that they meant to imply was that momentum was being lost somehow by the solar wind (or magnetosheath) plasma and communicated by an unspecified mechanism to force convection of the magnetospheric plasma (Hines, personal communication, 1985).

Based on observations, some discussed here, the latter is the correct choice for solar wind–magnetospheric interaction; the viscous effect is due to a boundary layer just inside the magnetopause. *Cole* [1961] pointed out that solar wind plasma flowing across the outer geomagnetic field will generate an electric field that would force convection of magnetospheric plasma, and would drive field-aligned and ionospheric currents. That the plasma can polarize effectively to maintain an electric field had already been explained by *Schmidt* [1960, 1979].

A magnetopause boundary layer of solar wind plasma has been found observationally, largely through the efforts of Eastman [*Eastman et al.*, 1976; *Eastman and Hones*, 1979] over a decade later (see his review in these Proceedings). This low latitude boundary layer (LLBL) could be the dynamo, but only if an efficient transfer process exists.

I have proposed a plasma transfer process that meets all requirements [*Heikkila*, 1982, 1984, 1986, 1997, 1998]; the process is reviewed briefly here. Basic concepts, Newton's laws and Maxwell's equations, are used to discuss the physical principles of solar wind–magnetospheric interaction. A model for the initial condition, in the equatorial plane, starts with a localized solenoidal current meander $\delta \mathbf{J}$, Figure 2(a), due to erosion.

Figure 2(b) shows the profile of the associated inductive electric field, by Lenz's law; an electromotive force (emf) $\varepsilon = \oint \mathbf{E} \cdot \mathbf{dl} = -d\Phi^M/dt$ is evident, where Φ^M is the magnetic flux through the chosen contour. Figure 1(a), in the meridian plane, is deficient in that it does not include an emf as an initial condition; Figure 1(b) does, in terms of the reversing E_t and Faraday's law curl $\mathbf{E} = -\partial \mathbf{B}/\partial t$.

There is observational evidence, both old and new, at the magnetopause as well as at low altitudes, to support this model. Some of these are reviewed here: electromotive force does exist at the magnetopause; closed field lines in the LLBL are a common feature; observations from low altitude spacecraft reveal the magnetopause current layer; and sporadic emission of keV electrons along B_n inside the magnetopause current sheet. The surprising conclusion is that reconnection occurs due a plasma transfer event.

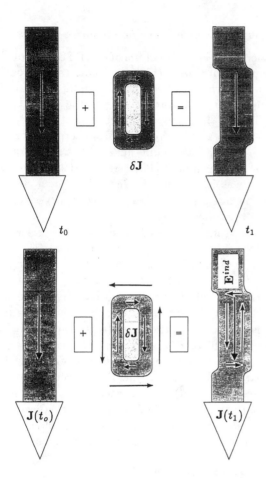

Figure 2. (a) A clockwise current perturbation $\delta \mathbf{J}$, in the equatorial plane, is needed to create more magnetic flux with a southward component, associated with erosion of the magnetopause. The inward meander of the current sheet at time t_1 is equivalent to the former current at time t_0 plus this perturbation loop. (b) By Lenz's law, an induction electric field $\mathbf{E}^{ind} = -\partial \mathbf{A}/\partial t$ in the counter-clockwise sense will oppose the current perturbation everywhere. The line integral $\varepsilon = \oint \mathbf{E} \cdot \mathbf{dl} = -d\Phi^M/dt$ around the perturbation is the electromotive force (emf). It is proposed that this induction electric field initiates a plasma transfer event.

2. MODEL FOR PLASMA TRANSFER

The initial condition for a plasma transfer event is a perturbation current $\delta \mathbf{J}$ shown in Figure 2(a), in the equatorial plane. The topology of the magnetic field is assumed to be like that in Figure 1; three dimensions are required to show both. The immediate cause is a pressure pulse from the magnetosheath, an inward push by the solar wind plasma associated with erosion, localized for causal reasons because energy cannot travel super-Alfvénically. The inductive electric field in Figure 2(b) is forced upon the plasma, not an electrostatic field; it is entirely local, opposed to the current perturbation

$\delta \mathbf{J}$. This process can happen anywhere on the magnetopause (dayside, flanks, lobes). It is also appropriate for the initiation of substorms in the inner magnetotail [*Heikkila et al.*, 2001], and for solar flares.

The plasma tries to respond to this onslaught by charge separation (polarization) but the response is hindered by the magnetic field. Because B_z is the dominant component of the magnetic field on either side of the magnetopause, the low (~ 0) Pedersen conductivity for a collisionless plasma in the tangential y direction limits polarization of charge in that direction, down or up on both sides of the current layer in Figure 2(b).

Such is the case at the upper and lower edges of the imposed current perturbation in Figure 2(b) in the case when the normal component of the magnetic field B_n vanishes (corresponding to a tangential discontinuity). Here too we must use the Pedersen conductivity. In other words, if $B_n \sim 0$ the plasma cannot respond at all to the imposition of the current meander by charge separation; the inductive electric field alone is the field that determines the motion of the plasma [*Heikkila*, 1982]. This is the low shear case of *Phan et al.* [1994].

On the other hand, with a finite B_n the plasma can polarize along it; the electron and ion mobilities are high along B_n so now we can use the very high direct conductivity. The plasma can try to cancel (or at least reduce) the normal component E_n (now $\sim E_{||}$).

An electrostatic field can have no effect on the electromotive force of the inductive field because its curl vanishes. Any reduction in the net $E_{||}$ in an arbitrary closed contour must involve enhancement of the perpendicular component $E_\perp$ at least somewhere, otherwise the curl (or emf) would be affected. The current carriers feel the electrostatic field caused by the charge polarization along B_n; the result is a tangential *electrostatic* field directed *oppositely* on the two sides of the localized current meander, as depicted in Figure 2(b). The high shear case of *Phan et al.* [1994] is the result.

Now we realize a pivotal result of this model: since both fields $\mathbf{E}$ and $\mathbf{B}$ reverse (approximately) for a high-shear magnetopause, their cross product $\mathbf{E} \times \mathbf{B}$ is unidirectional. Most of the plasma can cross the moving magnetopause as a fluid, in a dynamic process into the low latitude boundary layer, even onto closed magnetic field lines. There is now observational evidence of this: *Phan and Paschmann* [1996] have shown that the normal component of plasma velocity v_n is not affected for high shear, their Figure 7 [*Heikkila*, 1997].

The leading part of the perturbation (increasing current locally) acts as a dynamo with $\mathbf{E} \cdot \mathbf{J} < 0$, losing particles, momentum, and energy. The trailing part (decreasing current) acts as an electrical load, $\mathbf{E} \cdot \mathbf{J} > 0$,

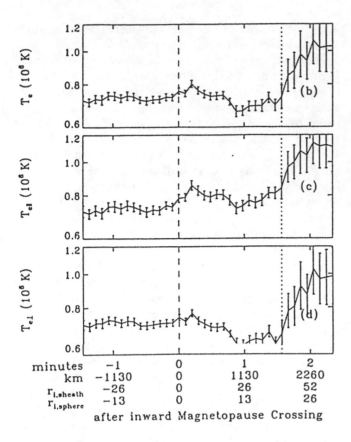

minutes   −1   0   1   2
km   −1130   0   1130   2260
$r_{i,sheath}$ −26   0   26   52
$r_{i,sphere}$ −13   0   13   26

after inward Magnetopause Crossing

Figure 3. A superposed epoch analysis of AMPTE data [*Phan and Paschmann*, 1996] of electron temperature data indicate that the temperature rose in one half of the current layer (mainly in $T_{\parallel}$), but that it decreased in the earthward half (mainly in $T_{\perp}$). Since electrons are one of the current carriers we can use T_e as an indicator of the electric field felt by all the current carriers. The conclusion is that the sense of the electric field reverses within the magnetopause current sheet. Curl **E** is finite, and an emf exists to convert stored magnetic energy into particle kinetic energy [*Heikkila*, 1997].

the particles gaining energy. The two are close together, as required, because energy cannot travel super-Alfvénically from dynamo to load.

Both volume integrals in Poynting's theorem are used, corresponding to electric and magnetic energies. These same terms have been used by *Lemaire and Roth* [1978] for impulsive penetration (IP), and by *Heikkila* [1982] for an analysis of plasma transfer (PTE), respectively.

3. OBSERVATIONS

There is sufficient observational evidence, both old and new, both at the magnetopause as well as at low altitudes, for validation of this model.

3.1. Electromotive Force at the Magnetopause

A superposed epoch analysis of AMPTE data of electron temperature data (Figure 3) was carried out by *Phan and Paschmann* [1996]. The results indicate that the temperature rose in one half of the current layer mainly in $T_{\parallel}$, thus supporting reconnection (see Figure 1(a) for the outflow region). However, T_e decreased in the earthward half mainly in $T_{\perp}$; that is contrary to the reconnection model [*Phan et al.*, 1996]. Since the discrepancy is in $T_{\perp}$ (not in $T_{\parallel}$) it is a local effect; it can not be due to the inflow of low energy electrons, e.g. from the ionosphere.

Since electrons are one of the current carriers we can use T_e as an indicator of the electric field felt by all the current carriers. We know that the magnetopause current is from dawn to dusk, a positive sign of **E · J** means that the electric field also has this dawn–dusk sense. But that is true only for a portion of the magnetopause current; for the inner (earthward) part T_e is reduced instead, relative to the magnetosheath. Since the magnetopause current must still be dawn–dusk (it still separates the magnetosheath magnetic field from the geomagnetic field) we can reach the conclusion that the sense of the electric field is in the dusk–dawn sense here, opposite to the outer part [*Heikkila*, 1997]. Curl **E** is finite, and an emf exists as in Figure 1(b) and 2(b).

3.2. The Transition Parameter

Figure 4 is called the transition parameter plot [*Hapgood and Bryant*, 1990; *Lockwood and Hapgood*, 1997]; it is based upon the idea that the LLBL is a mixture of magnetosheath and magnetospheric plasma. A scatter plot of plasma density against energy is made for a chosen interval. Higher plasma density implies a location close to the magnetopause, or even the magnetosheath, while a higher average energy (temperature) implies a location on inner edge of the boundary layer or the magnetosphere proper.

On Oct 17, 1992, GEOTAIL spacecraft skimmed the morningside magnetopause for several hours; multiple crossings of the current sheet make this orbit excellent for a study of plasma processes relevant to solar wind–magnetospheric interaction. Data from the Comprehensive Plasma Instrumentation (CPI) of the University of Iowa [*Frank et al.*, 1994] provided density, energy, temperature, velocity, and Alfvén velocity information for the entire 6 hour period [*Heikkila et al.*, 2002]. Plasma density in Figure 4(a), and the magnetic field in Figure 4(b) are plotted as a function of $T_{e\parallel}$ with no averaging (each point corresponds to an instrument cycle). At the left this figure shows the value appropriate to

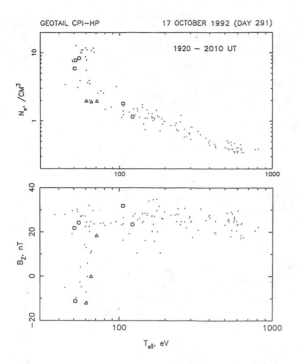

Figure 4.   (a) Data from the GEOTAIL satellite in a skimming orbit of the morning side magnetopause. This transition parameter plot is based upon the idea that the LLBL is a mixture of magnetosheath and magnetospheric plasma. A scatter plot of plasma density against energy is made for a chosen interval. (b) The magnetic field at the times when the data in (a) were taken. The special symbols correspond to certain times; the square to the right is at 1926.22 in Figure 8. These measurements show that magnetosheath plasma is able to traverse the magnetopause quite readily onto closed magnetic field lines.

the magnetosheath, while at the right are conditions in the LLBL or the outer magnetosphere.

The special symbols denote certain times are explained in the referenced article. For example, the open circle symbol to the right is the instrument cycle beginning at 1926:22 in Figure 8. The ratio is nearly unity over the entire energy range corresponding to bouncing particles between mirror points in the northern and southern hemispheres, therefore closed field lines. Of course some plasma is lost from the system due to the polarization current, and some along the field lines. These measurements show that solar wind plasma is able to traverse the magnetopause quite readily onto closed magnetic field lines in the LLBL.

3.3. Location of the First Open Field Line

The location of the first open field line on the dayside may be the deciding factor between the two alternatives of magnetic reconnection and a plasma transfer event. *McDiarmid et al.* [1976] established with the

Aloutte and ISIS satellites 3 decades ago that the first open field line is within the cusp (cleft) precipitation. Although the fluxes were low (as compared to the magnetosphere), with their high geometric factor they were able to detect the anisotropic pancake pitch angle distribution that is a mark of closed field lines well into the cleft. Figure 5 shows that

"On the average, anisotropic distributions extends about half way across the cleft region; however, it is significant that in some cases the pitch angle distribution is anisotropic throughout the entire cleft region."

This old result implies closed field lines within the cleft. Such observations have been observed repeatedly [e.g., *Mitchell et al.*, 1987; *Hall et al.*, 1991].

In contrast, reconnection with the model of Figure 1(a) with a constant electric field describes only an electric load with $\mathbf{E} \cdot \mathbf{J} > 0$. With a reversing magnetic field direction (south to north), the plasma flows into the reconnection region from both sides; the exit direction is to open field lines as shown [*Vasyliunas*, 1975; *Reiff et al.*, 1977; *Sonnerup et al.*, 1981; *Cowley*, 1982].

3.4. One Example of an Injection Event

Injection events observed on a Viking pass are shown in Figure 6 [*Woch and Lundin*, 1992]. The pass was

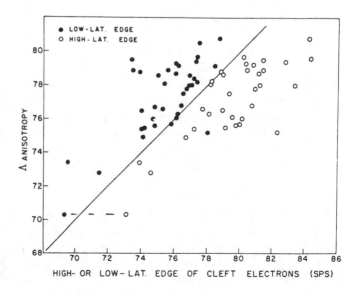

Figure 5. Plot of the highest latitude where anisotropic pitch angle distributions are observed against the high- and low-latitude edges of the cleft determined from the Soft Particle Spectrometer (SPS). Anisotropy implying close field lines extends about half way across the cleft region; however, it is significant that in some cases is anisotropic throughout the entire cleft region, [*McDiarmid et al.*, 1976].

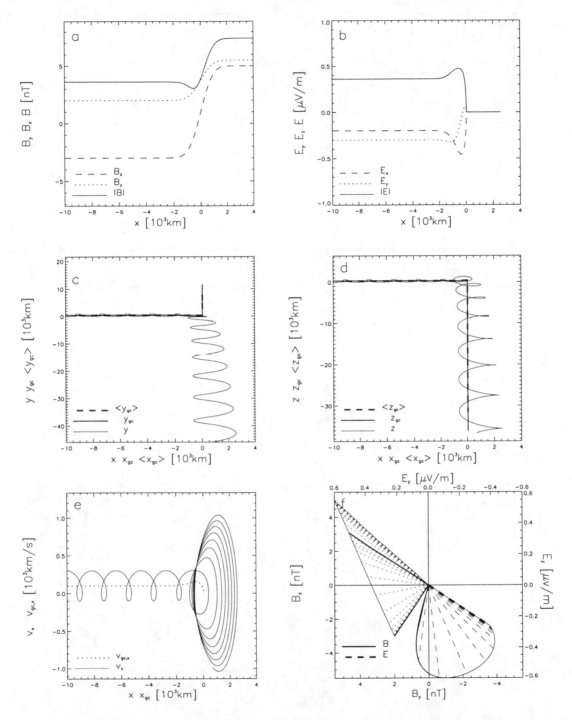

Figure 2. Trajectory of a proton in the "sheared" B-field distribution: *a*) Magnetic field; *b*) Electric field; *c*) xOy projection of particle (y), guiding center (y_{gc}), and averaged guiding center ($< y_{gc} >$) trajectories; *d*) xOz projections; *e*) particle's velocity (v_x) and guiding center velocity($v_{gc,x}$); *f*) hodograms of the B and E fields "seen" by the moving particle. The particle has not a large enough convection speed, $v_{cg,x}$, to overcome the magnetic potential energy. It is stopped in the region of large positive magnetic field gradient.

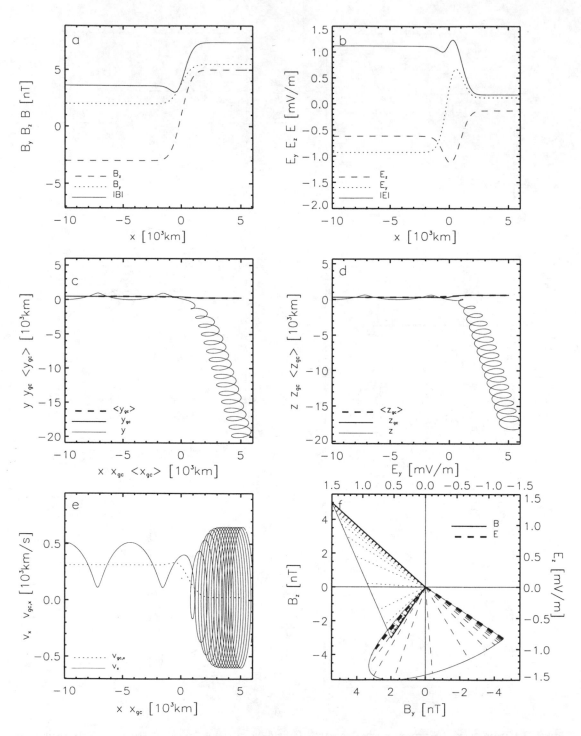

Figure 3. Same as figure **2** for a proton injected normal to a "sheared" B-field distribution. The scale length of the tangential discontinuity is comparable to the Larmor radius; the motion is not adiabatic. The particle has enough initial energy to cross the discontinuity. In this case the convection electric field is large enough to allow the proton to overcome the magnetic potential barrier and to penetrate into the magnetosphere.

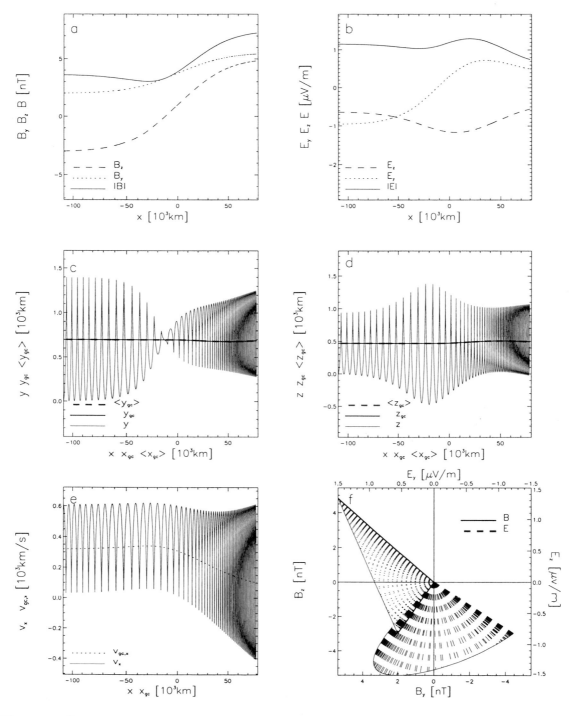

Figure 4. Same as figure 2 for a proton injected normal to a"sheared" B-field distribution whose scale length is here much larger than the Larmor radius. The guiding center path (thick lines, y_{gc}, in panels c-d) follows closely the particle trajectory, i.e. the solution of the equation of motion (9). Alfvén conditions are satisfied everywhere across the TD. The magnetic moment of the particle is adiabatically conserved.

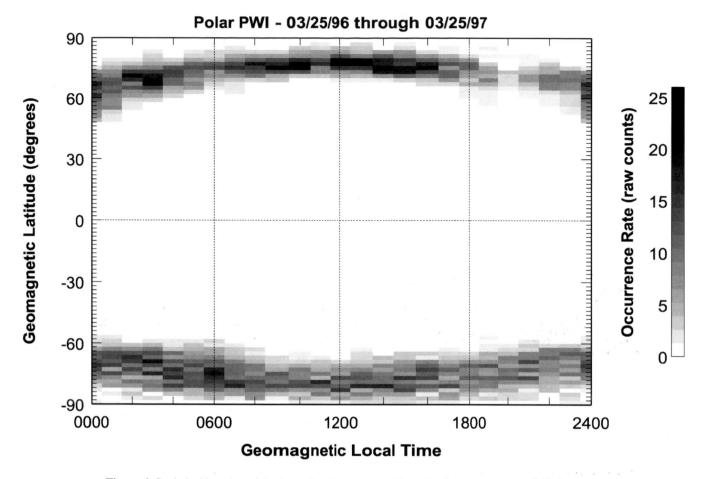

Figure 4. Statistical location of the boundary layer waves (from Polar). The distribution in positive latitudes is from the near-apogee data and that for negative latitudes is from the near-perigee data. The wave locations are centered at ~75° geomagnetic latitude at noon and at ~65° geomagnetic latitude at midnight. This is the same distribution as the Feldstein auroral oval.

kHz. The power for the other 19 frequency channels of electric (E) signals and 14 frequency channels of magnetic (B) signals have also been analyzed and are available. We show only the ~3 kHz E signals as a representative example.

As previously noted in Plate 1, in Anderson et al. (1982) and in Tsurutani et al. (1998b), the wave intensities vary from minute-to-minute, second-to-second and even millisecond-to-millisecond. Thus to try to understand local time dependences (and also the interplanetary control, to be discussed later), logarithms of the wave intensities were calculated and used for our analyses. The averages of the log values were then determined and used to construct the 10 min averages. The local time of the center of each 10 min wave interval was recorded and put into its proper bin. These values are plotted in Figure 5. The vertical bars indicate the standard deviation, σ. In these plots, 1 σ is approximately a factor of 10 in wave power. Since ± 2 σ cases exist for each hourly average value, there exists 10 min intervals where the wave log power aver-

age is ~100 times more intense and ~100 times less intense, or 4 orders of magnitude variation. Furthermore, within each 10 min interval (say the ± 2σ cases), there were further intensity variations. These features must be taken into account when considering wave-particle interactions and particle cross-field diffusion rates.

It should be noted that Polar crossed the LLBL magnetic field lines (where the waves were detected) at a variety of different distances from the Earth. For the near-apogee passes the range was ~5 R_e to ~8 R_e. Corrections for distance have not been made. However, it was noted in Tsurutani et al. (1998a) that there was little difference in wave intensity between Polar near-apogee and near-perigee passes, so at this time it is not certain what corrections (if any) should be made.

Examining the average wave power values in Figure 5, the peak values are found from 00-03 LT and 13-14 LT, just past midnight and local noon. Intensity minima are found from 05

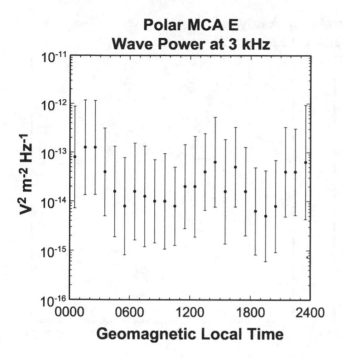

Figure 5. ~3 kHz electric wave intensity as a function of local time. This distribution was generated from one year of Polar near-apogee data. The solid dots are the log averages of 10 min intervals of wave intensities. The vertical bars represent one standard deviation in log averages. The maxima in wave intensities occur near midnight (1-2 LT) and near noon (13-14 LT) with minima near dawn and dusk. Because the BL waves are "broadbanded", these ~3 kHz electric waves are believed to be representative for other frequencies as well.

to 10 LT and from 18-21 LT, times near dawn and dusk, respectively.

The determination of Polar wave peak intensities occurring in the midnight sector is in agreement with the results of Gurnett and Frank (1978) from Hawkeye 1 (from 5.0 to 6.3 R_e) wave data analyses, but is in disagreement with the Akebono results given in Plate 2 (the wave peak intensities are in the noon sector). At the present time, we do not understand the meaning of the different results. One possibility may be the different period of analyses of the Akebono data (a whole solar cycle). Another possibility is the differences in the wave frequencies in the two studies (~3 kHz in Figure 5 and 5 Hz in Plate 2). These topics are currently being studied and will be reported in the near future.

Interplanetary Control

The interplanetary control of the wave intensity was examined by using time-lagged WIND spacecraft solar wind plasma and magnetic field data. We will discuss wave properties at four general local times: noon, dawn, dusk, and midnight.

Figure 6 gives the local noon (10 to 14 LT) 562 Hz electric wave intensity dependences on solar wind parameters. The waves have a slight dependence on the interplanetary B_Z (R = 0.23) and solar wind speed (R = 0.29). The wave intensities are largest during large IMF B_S (southward or negative B_Z) and the highest solar wind (V_{SW}) speeds.

Since magnetic merging at the magnetopause depends on both B_s and V_{SW}, one implication of these results is that the

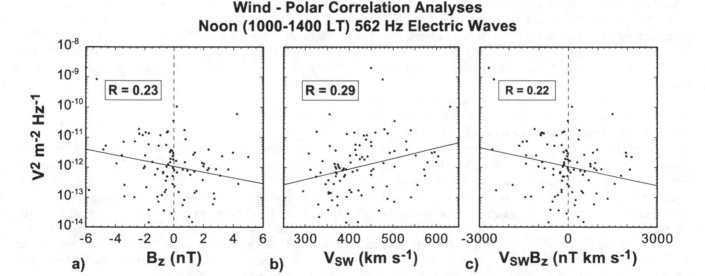

Figure 6. Correlation between time-lagged interplanetary WIND data and 562 Hz electric wave intensities for local dusk intervals. There is a slight correlation between solar wind velocity and IMF B_S with wave intensity.

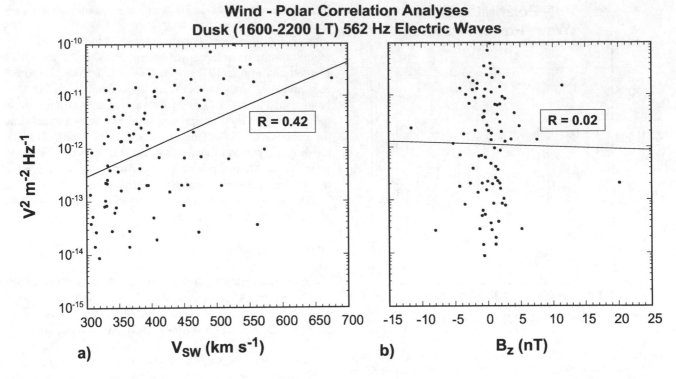

Figure 7. The relationship between time-lagged solar wind velocity and IMF B_Z and 562 Hz electric waves. There is a clear correlation with velocity (R = 0.42) while there is no apparent relationship between IMF B_Z (R = 0.02) and wave intensities.

wave intensities are dependent on the magnetic field merging rate. One possible scenario is that field-aligned currents are intensified during magnetic merging, and this in turn leads to greater wave intensities. In a previous LLBL study, Tsurutani et al. (1989) using ISEE-1 data found that there was only one interplanetary parameter that was correlated to wave intensities. The wave intensities were (slightly) higher when the IMF B_Z was directed southward, consistent with this present result.

Waves detected near local midnight were also found to have IMF B_Z dependences (with suitable delay times). This was shown in Tsurutani et al. (2001) and is not shown here to save space. The same arguments could be made for this interplanetary dependence. Dayside merging will be followed by nightside substorm-dependent magnetotail merging. The nightside merging will have the consequence of enhanced field-aligned currents and the generation of auroral zone plasma waves.

Figure 7 shows the primary dusk flank wave dependences. The wave intensities are strongly dependent on the solar wind velocity but not on IMF B_Z. As noted in Figure 5, the wave intensities are statistically lower than that for noon or for midnight. Thus one possible mechanism for wave generation at these local times is a Kelvin-Helmholz instability with the generation of large scale vortices (see article by Otto and Nykyri, this issue).

WAVE MODES

There are at least 4 electric wave modes and 3 electromagnetic wave modes present in the boundary layer. We will briefly discus the modes and their possible generation mechanism.

Electrostatic Modes

Lower hybrid resonance waves. Electric and magnetic waveforms observed in the range of 20 Hz to 25 kHz and the corresponding electric power spectra are shown in Figures 8 and 9, respectively. This data interval is taken at ~1332 UT, January 26, 1997. Polar was at 17.6 MLT, 4.9 R_e from the Earth, and at L = 6.7. This period was a geomagnetically active period with Kp = 4⁻. This wave interval occurred at the edge of an intense field-aligned current region, similar to that shown in Figure 3.

$E_{\parallel}$ and $B_{\parallel}$ correspond to the wave electric and magnetic field amplitudes along the ambient magnetic field. There are ~1.4 kHz waves present in both $E_{\parallel}$ and $E_{\perp}$. The electric waves

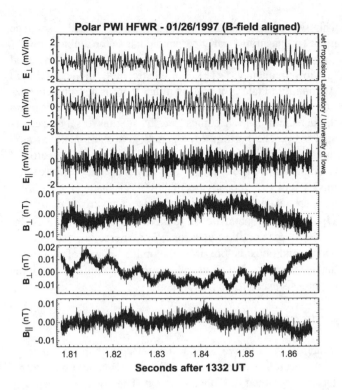

Figure 8. An example of showing the presence of ~1 kHz Polar "electrostatic" waves in the waveforms obtained in the frequency range of 20 Hz – 25 kHz. The electric components are shown in the upper 3 panels and the magnetic components in the lower 3 panels.

have peak amplitudes of $\sim \pm 1\,\mathrm{mV\,m}^{-1}$. The $E_{\parallel}$ component contains short duration packets of waves near f_{pe}. There are also ~150 Hz magnetic component waves with amplitudes of $\sim \pm 5 \times 10^{-3}\,\mathrm{nT}$. The latter waves are quasiperiodic.

In Figure 9, the lower hybrid frequency, f_{LH}, the electron cyclotron frequency, f_{ce}, and the electron plasma frequency, f_{pe}, are indicated. There is an uncertainty in the f_{pe} value due to an uncertainty in electron densities. The most probable value is used to calculate the f_{pe} value shown here (see discussion in Tsurutani et al., 2001). The time interval of analyses is 1332:01.8083 UT to 1332:01.865 UT.

The electric waves are present over a broad frequency range that extends from ~300 Hz to ~6 kHz. These $E_{\parallel}$ waves may be Doppler-shifted lower hybrid resonance (LH) waves. The greatest power is found near ~1 kHz when the majority of power is in the perpendicular (to the magnetic field) component ($\sim 3 \times 10^{-4}\,\mathrm{mV^2\,m^{-2}\,Hz^{-1}}$). At ~1 kHz, $E_{\parallel} \sim 5 \times 10^{-6}\,\mathrm{mV^2\,m^{-2}\,Hz^{-1}}$, over an order of magnitude lower. At ~5 kHz the situation is reversed. $E_{\parallel}$ is $10^{-4}\,\mathrm{mV^2\,m^{-2}\,Hz^{-1}}$ and $E_{\perp} \approx 2 \times 10^{-5}\,\mathrm{mV^2\,m^{-2}\,Hz^{-1}}$.

Electrostatic bipolar pulses, offset bipolar pulses and monopolar pulses. Onsets of bipolar structures have been recently detected for the first time (Tsurutani et al., 2001). An example can be found in Figure 10 at ~0827:08 UT on May 20, 1996. Polar was located at ~noon and at a magnetic latitude of 77°. The wave amplitudes start from noise and reach amplitudes of ~± 9 mV/m in the parallel component within ~0.05s, and are bipolar in form. The perpendicular component, monopolar in form, reaches amplitudes of ~± 2 mV/m. Note that there is no detectable magnetic field component present. However, it should be noted that the magnetic field variations onboard Polar are measured by search coil sensors. These sensors have a poor high-frequency response (see Tsurutani et al., 1998b) and thus very high frequency magnetic oscillations may have been missed.

Figure 11 shows an example of paired monopolar pulses. Polar was in the local noon sector. The two events from 0828:58.479 UT to 0828:58.482 UT and from 0828:58.490 UT to 08288:58.492 UT are examples where there is a negative pulse followed by a slightly delayed positive pulse. It has been argued that these offsets are evidence for dispersion within the bipolar pulses (Tsurutani et al., 1998b).

There is strong evidence that the bipolar pulses are Bernstein, Greene and Kruskal (BGK) (Bernstein et al., 1957) phase space electron holes (Omura, 1994; Ergun et al., 1999;

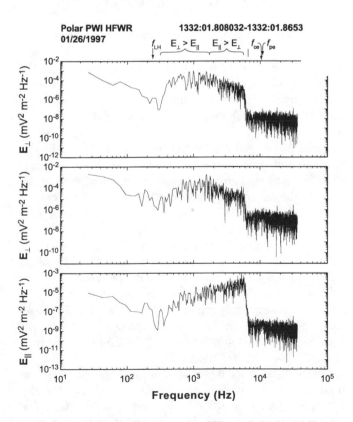

Figure 9. Power spectra for the waves of Figure 8. At f < 1 kHz, the wave perpendicular power is substantially greater than that aligned along B_0. However, at f > 5 kHz, the dominant power is found to be aligned with B_0.

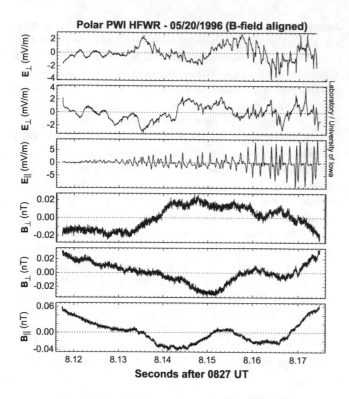

Figure 10. An onset of bipolar pulses (electron holes).

Goldman et al., 1999). Recent work has been performed investigating the degree to which ion acceleration can occur due to particle stochastic interaction with the paired monopolar structures (dispersed electron holes) (Muscietti et al., 2002).

Similar bipolar pulse structures are characteristic of low altitude auroral zone measurements taken on FAST (Ergun et al., 1999). The similarity of wave structures like these, in addition to the magnetic latitude location of the waves, enforce the idea that the LLBL is on auroral zone magnetic field lines.

Upper hybrid resonance frequency waves. Another electrostatic wave mode is shown in Figures 12 and 13. These are high-frequency modes often found in the presence of ~1 kHz LHR waves. In Figure 12, it can be noted that the ~1 kHz waves are continuously present and that the ~10 kHz waves occur sporadically. The latter waves are noted in the $E_\perp$ and $E_\parallel$ channels from ~1330:57.530 to 1330:57.540 UT and from ~1330:57.544 to 1330:57.558 UT.

Figure 13 shows the power spectra for the event. Polar was in the dusk sector, at ~5 R_e distance and L ~ 6.8. A clear, narrow peak is noted at ~10 kHz. However, an interesting feature is that the wave power in $E_\perp$ is almost an order of magnitude larger than that in $E_\parallel$. This fact rules out the possibility of these waves are longitudinally polarized Langmuir waves.

This wave mode is most likely the upper hybrid resonance mode, propagating nearly perpendicular to the ambient field.

Ion acoustic waves. Kasahara et al. (2001) have analyzed the low-frequency (<1 kHz) electrostatic modes onboard Akebono. In Plate 3, it is shown that there is a strong correlation between this intense broadband electrostatic noise and the detection of "ion conics" or TAI (Transversely Accelerated Ions). The bottom three panels give the ion fluxes as a function of energy and time. There is an excellent temporal relationship between the intense low-frequency electrostatic modes and the appearance of TAI. Kasahara et al. (2001) have speculated that because the noise is polarized parallel to B_0, it is an ion acoustic wave mode. This mode can be generated by precipitating electrons. The interaction between the waves and ions is possibly the main source of ion heating/acceleration. Detailed calculations to confirm/deny this hypothesis needs to be performed.

Electromagnetic Waves

Kinetic Alfvén waves. At the lowest end of the wave magnetic spectrum, there are broadband magnetic fluctuations. This can be noted in Plate 3, panel b. The power is greatest at frequencies below ~25 Hz. The local proton cyclotron frequency, f_{pc} is ~100 Hz. There has been some debate as to the nature of these fluctuations which are not well understood. Are they ion cyclotron waves, kinetic Alfvén waves or turbulence? Satellite passage over filamentary current structures (shown in Figure 3) will introduce power into the magnetic sensors, thus some of this "power" may not be true wave power. Tsurutani and Thorne (1982) and Gendrin (1983) have shown that some of the low frequency LLBL wave

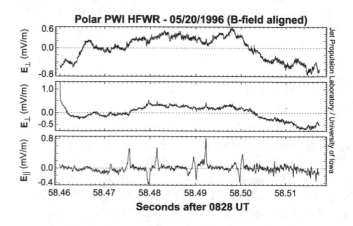

Figure 11. An example of paired monopolar pulses. The electric field pulses are aligned along B_0. Paired monopolar pulses have been speculated to be dispersed bipolar pulses (see Tsurutani et al., 1998b).

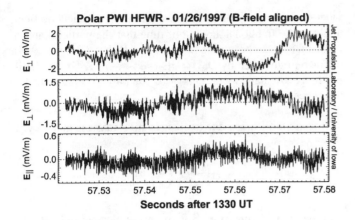

Figure 12. ~10 kHz electric waves superposed on top of ~7 × 10² to 3 × 10³ Hz electric waves observed in the waveforms obtained in the frequency range of 20 Hz – 25 kHz.

power may be electromagnetic ion cyclotron waves. Stasiewicz et al. (2001) studying waves in the region of the magnetopause, have shown that some large scale features have E/B ratios close to the Alfvén speed. They argue that these are nonlinear Alfvén waves. For smaller scale structures, this E/B ratio increases, consistent with the presence of kinetic Alfvén waves.

kHz Whistler mode waves. Figure 14 shows ~0.01s bursts of electromagnetic whistler mode waves occurring at ~4.9 kHz on May 20, 1996. Polar was located at 6.1 R_e, 80° invariant latitude and at ~1148 LT. The inset shows that the wave trains are quite coherent. The local magnetic field was ~197 nT and thus the electron cyclotron frequency was 5.5 kHz. The peak wave amplitude was ~5 × 10⁻³ nT in the $B_\perp$ component. These intensities would lead to only very weak pitch angle diffusion and would not cause rapid disruption of any electron beam generating the waves.

It has been shown by Tsurutani et al. (1998b, 2001) that the 1-5 kHz whistler mode wave can be generated by cyclotron resonance with auroral zone electrons. For this case, resonant electron parallel energies would be of the order of ~100 eV to ~1.1 keV (for 1 kHz and 5 kHz waves, respectively).

These electromagnetic waves may be the same phenomenon as "VLF hiss" detected at lower altitudes (Barrington et al., 1971; Hoffman and Laaspere, 1972). The latter work noted a close association between hiss and auroral 100-eV to several keV electrons, in agreement with the above cyclotron resonance calculations.

200 Hz electromagnetic waves. The presence of low-frequency ~200 Hz electromagnetic waves is a typical feature of the Polar apogee pass data. An example taken in the dawn sector is shown in Figure 15 on December 16, 1996. Polar was 8.7 R_e from the Earth, at ~0300 LT and at 81° invariant latitude. The wave amplitude is ~3 x 10⁻² nT peak-to-peak,

with the dominant power primarily in the $B_\perp$ component. The $B_\parallel$ component was smaller, with a value ~1.0 × 10⁻² nT. These waves may be generated by field-aligned currents and/or electron temperature anisotropies driving the electromagnetic lower hybrid instability (Lakhina, 1980).

WAVE GENERATION MECHANISMS

A linear coupled velocity shear-lower hybrid instability model has recently been developed by Lakhina and Tsurutani (1999) for the generation of the broadband plasma waves. The model is based on two-fluid equations and is fully electromagnetic and takes into account the free energy present in the boundary layer: field-aligned currents, gradients in the currents, and gradients in plasma densities and magnetic fields. The dispersion relationship generalizes the dispersion relations for several plasma modes, including the lower hybrid instability (Gary and Eastman, 1979), the modified two stream instability (Lakhina and Sen, 1973), beam modes (Dum, 1989; Lakhina, 1993), and current convective and

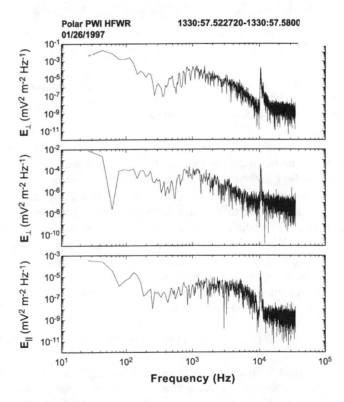

Figure 13. The power spectra for waves in Figure 12. At ~1 kHz, the power in $E_\perp$ (~10⁻⁴ mV² m⁻² Hz⁻¹) is almost 2 orders of magnitude larger than that of $E_\parallel$ (~10⁻⁶ mV² m⁻² Hz⁻¹). However at f = 5 × 10³ Hz, the $E_\parallel$ power is larger (~2 × 10⁻⁶ mV² m⁻² Hz⁻¹ versus ~1 × 10⁻⁶ mV² m⁻² Hz⁻¹). The emission at ~1 × 10⁴ Hz is polarized transverse to B_0. $E_\perp$ = 1-2 × 10⁻⁴ mV² Hz⁻¹ and $E_\parallel$ = 2 × 10⁻⁵ mV² m⁻² Hz⁻¹

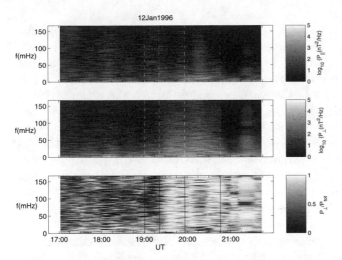

12Jan1996

Figure 2. Wave power spectra for the WIND crossing of the magnetopause shown in Fig. 1. $P_\parallel$, $P_\perp$ and P_{tot} are the power spectral densities obtained from $|\delta B_\parallel|^2$, $|\delta B_\perp|^2$ and $|\delta \mathbf{B}|^2$ respectively. Notice that prior to crossing the magnetopause wave power is primarily compressional, coincident with the magnetopause crossing (indicated by the vertical line), the wave activity is primarily transverse. Note that the compressional wave component remains essentially the same before and during the magnetopause crossing.

of the magnetosheath) of the transverse spectrum falls off as frequency increases.

In fact, ultra-low frequency (ULF) waves (with frequencies below 500 mHz) dominate the spectrum of nearly every magnetopause crossing (*Perraut et al.*, 1979; *Rezeau et al.*, 1993; *Song et al.*, 1993c; *Song*, 1994; *Phan and Paschmann*, 1996, and references therein). It has been suggested that these waves are associated with mode conversion of MHD waves in the magnetosheath to kinetic Alfvén waves (KAWs) at the magnetopause near a field line resonance location (*Lee et al.*, 1994; *Belmont et al.*, 1995; *De Keyser et al.*, 1999). The mode conversion process can explain (a) a change in wave polarization at the magnetopause and (b) the amplification of the transverse magnetic field component by an order of magnitude (*Johnson and Cheng*, 1997b).

Because the interplanetary magnetic field changes orientation frequently, it is possible to characterize the magnetopause wave activity as a function of magnetic shear (defined to be the angle between the magnetic field in the magnetosheath and the magnetic field on the magnetospheric side of the magnetopause). Because the mode conversion process exhibits a strong dependence on this magnetic field rotation theory/data comparisons of expected wave signatures can be used effectively as a test of the mode conversion process. Data for the comparison was taken from the ISEE1,

ISEE2 and WIND spacecrafts. Thirteen magnetopause crossings were considered where sizeable compressional magnetic field fluctuations were found in the magnetosheath. These cases provided coverage of the magnetic rotation angle from 0 to 180 degrees. Windowed magnetic power spectra were obtained for parallel and transverse magnetic fluctuations during each crossing. The magnetosheath and magnetopause spectra were compared for each crossing to quantify the wave amplification for each crossing. The wave amplification factor, $P_{\perp mp}/P_{\perp msh}$, where $P_{\perp msh}$ and $P_{\perp mp}$ refer to average values of the power spectral density of $\delta \mathbf{B}_\perp$ in the magnetosheath and magnetopause respectively, is shown in Fig. 3 for these magnetopause crossings for frequencies of 25 and 50 mHz. From the study (*Johnson and Cheng*, 2001), it was deduced that (1) the transverse wave component at the magnetopause is not significantly amplified below a threshold magnetic shear angle (approximately 50 degrees), (2) greatest amplification is for magnetic shear between 70 and 180 degrees, and (3) waves with higher frequencies are less amplified.

3. KINETIC ALFVÉN WAVES AT THE MAGNETOPAUSE

These observations can be understood in the context of resonant mode conversion of compressional Alfvén waves into kinetic Alfvén waves (KAW) (*Hasegawa*, 1976) at the magnetopause. Resonant mode conversion occurs when a compressional Alfvén wave propagates into a region with gradients in $k_\parallel V_A$ such as at the magnetopause where the Alfvén velocity can increase by at least a factor of 10. At the magnetopause, gradients in the direction normal to the magnetopause boundary are dominant compared with gradients along the boundary, and we can approximate the background plasma and magnetic field profiles as functions of the coordinate, x, in the direction normal to the magnetopause. We assume the magnetic field is of the form

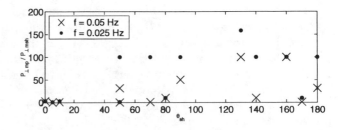

Figure 3. Amplification ($P_{\perp mp}/P_{\perp msh}$) of waves at the magnetopause as a function of magnetic shear.

$\mathbf{B} = B_0(x)\mathbf{b}$ where $\mathbf{b} = \cos\theta_b(x)\hat{\mathbf{z}} + \sin\theta_b(x)\hat{\mathbf{y}}$ and the magnetic field angle, θ_b rotates by an angle θ_{sh} across the magnetopause. The equilibrium profiles vary smoothly across the magnetopause on a scale of 10 ion gyroradii (ρ_i). For such a configuration, wave propagation is well-described by the kinetic-fluid model (*Cheng and Johnson*, 1999) which simplifies to the following set of dimensionless coupled equations for $\mathcal{W}_\parallel = \delta p + B_0 \delta B_\parallel$ and $\mathcal{W}_x = i B_0 \delta B_x$.

$$\frac{d^2\mathcal{W}_\parallel}{dx^2} = -(k_A^2 - k_S^2)\mathcal{W}_\parallel + (k_A^2 - k_\parallel^2)\delta p$$

$$+ k_\parallel \left[\frac{d}{dx} \log\left(\frac{k_\parallel^2}{k_A^2}\right) \right]\mathcal{W}_x \qquad (1)$$

$$\mathcal{K}\mathcal{W}_x \equiv \left[1 + \frac{T_e}{T_i}(1 + \hat{\eta}\frac{Z_i'}{2})(\frac{-2}{Z_e'})\right]k_\parallel^2\rho_i^2\frac{d^2\mathcal{W}_x}{dx^2}$$

$$= \left(k_\parallel^2 - k_A^2\right)\mathcal{W}_x - k_\parallel\frac{d\mathcal{W}_\parallel}{dx} \qquad (2)$$

$$\hat{\eta}\mathcal{W}_x \approx \int dx'\frac{dk_x}{2\pi}e^{ik_x(x-x')}b\frac{(\Gamma_0(b) - \Gamma_1(b))}{(1 - \Gamma_0(b))}\mathcal{W}_x(x') \qquad (3)$$

$\mathbf{k}_S$ is the wavevector in the plane perpendicular to x; $k_\parallel = \mathbf{k} \cdot \mathbf{b} = \mathbf{k}_S \cdot \mathbf{b} = k_S \cos\theta_{sb}$, where θ_{sb} is the angle between $\mathbf{b}$ and $\mathbf{k}_S$; and $k_A^2 = \omega^2/V_A^2$ is the Alfvén wavevector where ω is the wave frequency and V_A is the Alfvén velocity. Z_s' is the derivative of the plasma dispersion function of argument $\zeta_s = \omega/\sqrt{2}k_\parallel v_{ts}$ for species s with thermal velocity v_{ts}. We have taken the plasma to be isotropic. The operator $\hat{\eta}$ is a weakly nonlocal operator introduced by finite Larmor radius effects. The integration involves $\Gamma_n(b) \equiv I_n(b)e^{-b}$ where $b \equiv (k_x^2 + k_S^2 \sin^2\theta_b)\rho_i^2$. Near the location where $k_\parallel^2 = k_A^2$, $\hat{\eta} \approx 1 + \mathcal{O}(\rho_i^2 d^2/dx^2)$ and for $k_A^2 \gg k_\parallel^2$ (as occurs when $k_\parallel$ is small), the contribution of $\hat{\eta}$ vanishes.

The pressure equation required for the compressional wave is, $\delta p \approx (1 - 1/\tau)\mathcal{W}_\parallel$ where $\tau = 1 + \sum_s \beta_s(1 + Z_s'/2)$ and summation is over all species, s. The function $1/\tau$ is well behaved for the frequencies of interest (in contrast to the MHD approach which gives a singularity where $\omega^2 = k_\parallel^2 C_S^2/(1 + \beta)$) (*De Keyser et al.*, 1999). In a cold, isotropic plasma $\zeta_i \gg 1$, $\tau \to 1$ and there is no contribution from this term. In a warm plasma with $\zeta \sim 1$, $\tau \sim \mathcal{O}(1)$ and this term only introduces weak damping to the compressional wave (*Johnson and Cheng*, 1997a), so the sound resonance is not very important. Moreover, near the Alfvén resonance where $k_\parallel^2 = k_A^2$, the pressure term vanishes from Eq. 1, and Larmor radius corrections in the term proportional to δp are not critical for describing wave behavior near the Alfvén resonance.

Generally, the Alfvén velocity increases across the magnetopause from the magnetosheath side so that k_A^2 is a monotonically decreasing function. In the magnetosheath the wave is propagating which requires $k_A^2 > k_S^2$. As the wave propagates across the magnetopause, k_A decreases until $k_A^2 = k_S^2$ where the compressional wave is cutoff. Beyond that location, the compressional wave decays. However, at the location $k_A^2 = k_\parallel^2$ the decaying compressional wave encounters the Alfvén resonance where it can be reflected out of phase from the incoming wave. Near the resonance location the parallel magnetic field is well behaved, but the transverse fields are singular.

Ion gyroradius effects resolve the singular behavior and are described by the kinetic response integral operator, $\mathcal{K}$ (*Cheng and Johnson*, 1999). The model includes the effects of the parallel electric field through the quasineutrality condition, is valid for both large and small $k_\perp^2\rho_i^2$, and reduces to the KAW and inertial Alfvén wave dispersion relations in the appropriate limits. (Note that near the location where $k_\parallel \to 0$, the wave enters the inertial regime and decays on the scale of the electron skin depth (*Cheng and Johnson*, 1999)).

We solve Eqs. 1 and 2 numerically on a nonuniform discrete grid and obtain the solutions through matrix manipulation. Boundary conditions are imposed at the magnetosheath and magnetosphere boundaries. The boundary condition in the magnetosheath is an incoming compressional MHD wave. At the magnetosphere boundary, the compressional MHD wave is decaying. For the KAW only radiating/decaying solutions are allowed. Boundary conditions are imposed asymptotically. The Alfven velocity is taken to increase by a factor of 10 across the magnetopause and the magnetic field rotates through an angle, θ_{sh}.

A good measure of the efficiency of mode conversion at the magnetopause is the amount of compressional wave absorption in the magnetopause layer. Energy absorption is determined by comparing the Poynting flux ($\delta\mathbf{E} \times \delta\mathbf{B} \cdot \hat{\mathbf{x}}$) of the incident compressional wave (S_I) with the Poynting flux of the reflected (S_R) and transmitted (S_T) compressional waves. The Poynting flux of the KAW in the magnetopause near the mode conversion layer is $S_{KAW} = S_I + S_R - S_T$. In the magnetopause, the transverse magnetic field component is mainly from the KAW, therefore $S_{KAW} \sim P_{\perp mp}$, and the compressional wave absorption, $A \equiv (S_I + S_R - S_T)/S_I$ is proportional to the wave amplification, $P_{\perp mp}/P_{\perp msh}$. Depending on the profiles of V_A and $\mathbf{k}_S \cdot \mathbf{b}$, there can be up to three resonance locations in the magnetopause. The absorbed energy is converted to KAWs which: (a) propagate back into the magnetosheath (one resonance location), (b) propagate into both magnetosheath and magnetosphere (two resonance locations), or (c) couple to a quasi-trapped kinetic Alfvén wave (three resonance locations).

To determine the total absorption as a function of frequency and magnetic shear, we sum the absorption over the wavevector spectrum of incoming compressional waves. To do this, we assume that all wavevectors lie on a dispersion surface in wavevector space defined by $\omega(\mathbf{k}) = \text{const}$ and integrate over the dispersion surface. We integrate over the angles of $\mathbf{k}$: θ_{k0} is the angle between k and the magnetic field in the magnetosheath and ϕ_{k0} is the azimuthal angle in planes perpendicular to the magnetosheath magnetic field, $\mathbf{B}_{msh}$. For compressional waves the dispersion surface is approximately defined by $\omega^2 \approx k^2(V_A^2 + C_s^2 \sin^2 \theta_{k0})$. The wave vectors are approximately distributed on an ellipsoid with major radius $k = k_A$ and minor radii $k = k_A/\sqrt{1 + C_s^2/V_A^2}$. The absorption spectrum as a function of frequency is obtained by integrating over the ellipsoid—that is, over the angles (θ_{k0}, ϕ_{k0}) with $k^2 \approx k_A^2/(1 + C_s^2 \sin^2 \theta_{k0}/V_A^2)$ imposed by the dispersion relation. Moreover, compressional waves typically have $k_\perp \gg k_\parallel$ so it is reasonable to assume that the spectrum is highly peaked around $\theta_{k0} = \pi/2$. On the other hand, there is no compelling reason to expect that the initial wave spectrum depends on the direction ϕ_{k0}.

The wave observations show a strong dependence of amplification ($P_{\perp mp}/P_{\perp msh}$) on the shear angle across the magnetopause. For small shear angles, there is little wave amplification, while above a threshold amplification is enhanced and relatively level. The minimum in amplification for small shear is consistent with the mode conversion mechanism because the waves in the magnetosheath propagate nearly perpendicular to the magnetic field. The absorption coefficient, $A(\omega, \theta_{sh})$ is presented in Figure 4. The absorption is obtained by computing the absorption coefficient as a function of $(\omega, \theta_{sh}, \theta_{k0}, \phi_{k0})$ and performing an integration over the variables (θ_{k0}, ϕ_{k0}) with uniform weight in ϕ_{k0} and a strongly peaked weighting function about $\theta_{k0} = \pi/2$.

The absorption is the result of mode conversion to KAWs and measures the efficiency of the mode conversion mechanism. The absorbed energy is the Poynting flux of the KAW which radiates away from the mode conversion location. The Poynting fluxes scale as the group velocity multiplied by spectral density. Because the KAW radiates slowly across the magnetic field, its amplitude must be greatly increased compared with the amplitude of the incoming MHD wave in order to carry away the mode converted energy from the field line resonance location. For KAWs the Poynting flux

is approximately, $S_{KAW} \sim (\omega/k_x)k_x^2\rho^2/(1 + k_x^2\rho^2)P_\perp$, while for the MHD wave $S_{MHD} \sim V_A P_\perp$. The KAW wavevector, k_x can be estimated from dominant balance in Eq. 2— $k_x \sim (\rho_i^2 L)^{-1/3}$ where L is the scale length of the Alfvén velocity gradient at the magnetopause. One can then estimate from the linear dispersion relation that the wave amplification $P_{\perp KAW}/P_{\perp MHD} \sim A(1 + (\rho/L)^{2/3})V_A/[2\pi f\rho(\rho/L)^{1/3}]$. For typical magnetopause parameters: $V_A \sim 300$ km/s, $\rho_i \sim 50$ km, $L \sim 500$ km, $f = 25$ mHz, $P_{\perp KAW}/P_{\perp MHD} \sim 100A$.

Because $|\delta\mathbf{B}_\perp^2|$ amplification scales directly with compressional wave absorption, the results of Figure 4 can be compared qualitatively with observed $|\delta\mathbf{B}_\perp^2|$ amplification. The important features to notice are: (1) for angles greater than 50° the absorption is approximately constant, but for smaller shear there is a trough in A and (2) the absorption decreases weakly as frequency increases for angles larger than 50°. However, for angles less than 50° there is a significant broadening of the trough for higher frequency with far less absorption. These qualitative properties correspond well to observations of $|\delta\mathbf{B}_\perp^2|$ amplification as a function of magnetic shear angle and frequency as discussed in Fig. 3. The quantitative differences between the theory and data (for example, the theoretical threshold angle is smaller) can be attributed to the uncertainly involved in analyzing the data and the simplifications of the theoretical model.

4. Plasma Heating and Particle Transport Associated with Kinetic Alfvén Waves

Large amplitude transverse wave activity accompanies nearly every magnetopause crossing (*Perraut et al.,*

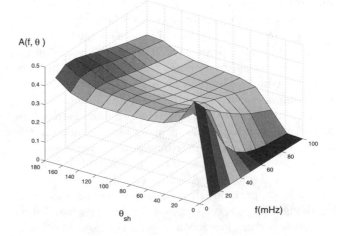

Figure 4. Absorption coefficient as a function of frequency and magnetic shear angle θ_{sh} (degrees).

1979; *Rezeau et al.*, 1993; *Song et al.*, 1993c; *Song*, 1994; *Phan and Paschmann*, 1996; *Johnson et al.*, 2001).Often the observed wave amplitudes at the magnetopause can be significantly large compared with the background magnetic field, B_0. It is not uncommon for $\delta B_\perp / B_0$ to be the order of 0.2 or even larger (*Rezeau et al.*, 1993; *Song et al.*, 1993c,b). Moreover, in the cusp, it is likely that $\delta B_\perp / B_0 \sim 1$ because of the weak background field.

The magnetopause provides a boundary between magnetosheath plasma and magnetospheric plasma. Plasma that leaks into the boundary layer is often found to have distinctive particle distributions indicative of acceleration processes. For example, electron distributions in the boundary layer are often found to be accelerated in the direction parallel to the magnetic field. These distributions have been identified as signatures of reconnection events, but the process which accelerates the electrons is not well known. It was suggested that electrons could also be accelerated by kinetic Alfvén waves (*Hasegawa and Chen*, 1976; *Lee et al.*, 1994) which preferentially heat electrons in the parallel direction due to the parallel electric field. Thermal electrons trapped in the wave potential would be heated leading to a slight increase in the parallel direction consistent with observations.

On the other hand, ions in the sheath transition layer and boundary layers often exhibit significant anisotropy with $T_\perp > T_\parallel$ (*Anderson et al.*, 1991; *Song et al.*, 1993a). Moreover, *Wilber et al.* (2001) have reported unusual low energy ion distribution components in the low-latitude boundary layer with pitch angles intermediate between 0 and 90 degrees observed by WIND/3DP during equatorial passes. The particles appeared to have undergone adiabatic streaming from a stronger magnetic field region with heating occuring near the magnetopause as deduced from the mirror ratios. The low energy ions appear to have been heated perpendicular to the magnetic field and in some events the core of the distribution appears to be flattened. The ion distributions presented by *Song et al.* (1993a) also illustrate that in the inner boundary layer the slope of the low energy component of the ion distribution function is flattened compared with the magnetosheath distribution suggestive that a physical process may be heating the low energy core of the distribution to higher energies.

In a recent study *Johnson and Cheng* (1997b) showed that when the kinetic Alfvén wave amplitude is sufficiently large, particle orbits can become stochastic leading to significant transport and particle heating. When the orbits became stochastic in an inhomogeneous background field, the particles could diffuse rapidly across

the magentic field with a diffusion coefficient of $D \sim 10^9 m^2/s$.

In this paper we will concentrate on particle heating. To study ion heating in the presence of a kinetic Alfvén wave, we will prescribe the electromagnetic fields consistent with the kinetic Alfvén waves. We investigate the particle motion in those prescribed fields as a function of the wave amplitude. The study will consist of a sequence of Poincaré sections taken at different wave amplitude which demonstrate the onset of stochasticity. The results will demonstrate: (1) stochastic ion heating can result through nonlinear coupling between low frequency waves and cyclotron motion, (2) ions can be heated transverse to the magnetic field leading to temperature anisotropy ($T_\perp > T_\parallel$) as observed at the magnetopause (*Wilber et al.*, 2001), and (3) the stochastic process will deplete the core of the ion distribution function leading to a flattened core of the distribution function similar to the observations of (*Wilber et al.*, 2001; *Song et al.*, 1993a).

5. KINETIC ALFVÉN WAVES AND ION HEATING

The kinetic Alfvén wave (*Hasegawa*, 1976) is well described by three scalar quantities—ϕ, $A_\parallel$, and $\delta B_\parallel$ (*Cheng and Johnson*, 1999). The fields associated with the kinetic Alfvén wave are obtained through Maxwell's equations given in Gaussian units.

$$\mathbf{E} = -\nabla\phi - \frac{1}{c}\frac{\partial A_\parallel \mathbf{b}}{\partial t} \tag{4}$$

$$\mathbf{B} = \mathbf{B_0(x)} + \delta\mathbf{B} \approx (B_0(x) + \delta B_\parallel)\mathbf{b} - \mathbf{b} \times \nabla A_\parallel \tag{5}$$

where $\mathbf{b}$ is the unit vector in the direction of the magnetic field. The vector potential is related to the electrostatic potential by introduction of a secondary potential, ψ defined by

$$E_\parallel = -\nabla\psi = -\nabla_\parallel\phi - \frac{1}{c}\frac{\partial A_\parallel}{\partial t} \tag{6}$$

The fields are obtained by solving the set of equations described in *Cheng and Johnson* (1999) for the prescribed background magnetic field. If the background field is uniform, the kinetic Alfvén wave is a simple sinusoidal with $\phi = \phi_0 \cos(\mathbf{k} \cdot \mathbf{x} - \omega t)$, $\psi \approx -(T_e/T_i)k_\perp^2 \rho_i^2 \phi/(1 + k_\perp^2 \rho_i^2)$, $A_\parallel = (k_\parallel/\omega)(\phi - \psi)$. The wave satisfies the approximate dispersion relation

$$\omega^2 = k_\parallel^2 V_A^2 \left(1 + (1 + \frac{T_e}{T_i})k_\perp^2 \rho_i^2\right) \tag{7}$$

where we ignore damping and take a Padé approximation for the Bessel function as described in *Cheng and*

Johnson (1999). The kinetic Alfvén wave is incompressible at low β so that $\delta B_\parallel \approx 0$, but at $\beta \sim 1$ and short wavelength, $\delta B_\parallel$ may also included as prescribed in *Cheng and Johnson* (1999). For the chosen parameters ($k_\perp \rho_i = 3$, $k_\parallel \rho_i = 0.05$, and $\beta = 1$), $|\delta B_\parallel| \sim 0.55|\delta B_\perp|$ and is 90° out of phase.

Particle orbits are determined by the equation of motion

$$m_i \frac{d^2 \mathbf{r}}{dt} = q_i (\mathbf{E} + \frac{\mathbf{v}}{c} \times \mathbf{B}) \qquad (8)$$

which we normalize to

$$\ddot{\mathbf{X}} = -\tilde{\nabla}\Phi + \dot{\mathbf{X}} \times (\Delta \hat{\mathbf{b}} + \frac{\delta \mathbf{B}}{\tilde{B}_0}) \qquad (9)$$

where $\mathbf{X} = \mathbf{r}/\rho_i$, $\tau = \Omega t$, $\tilde{\nabla} = \rho_i \nabla$, $\Phi = q_i \phi / T_i$, with $\Omega = q_i \tilde{B}_0 / m_i c$, $\rho_i = \sqrt{T_i/m_i}/\Omega$, $\Delta = |B_0(x)/\tilde{B}_0|$ and $\dot{\mathbf{X}} \equiv d\mathbf{X}/d\tau$, and $\tilde{B}_0$ is a characteristic value of magnetic field.

To investigate the behavior of particles in the kinetic Alfvén waves, we plot Poincaré sections for particle orbits. This technique is standard and has been applied to electrostatic waves to understand plasma heating well above the cyclotron frequency (*Karney and Bers*, 1977), at near the cyclotron frequency (*Hsu et al.*, 1979). Points on Poincaré sections are plotted at constant particle gyrophase, γ, with the requirement that $\dot{\gamma} < 0$. In the absence of waves, this would correspond with one point per gyroperiod. At each crossing of the phase space plane defined by $\gamma = 0$ where $\mathbf{k} \cdot \mathbf{b} \times \dot{\mathbf{X}} = k_\perp v_\perp \sin(\gamma)$, we plot the value of the magnetic moment, $\mu \equiv |\mathbf{b} \times \dot{\mathbf{X}}|^2/\Delta$, versus $\Psi = \mathbf{k} \cdot \mathbf{x} - \omega t$, taken modulo 2π. In the following plots, at least 1000 points are taken for each trajectory to resolve the phase space structure (more near a separatrix). The choice of these variables is good for examining particle heating in the presence of the wave because the magnetic moment is an adiabatic invariant which follows well defined trajectories in phase space. The wave phase is an obvious choice because there is a direct correlation between the adiabatic particle motion and the wave amplitude.

To examine the onset of stochastic particle behavior in the presence of large amplitude kinetic Alfvén waves, we examine a sequence of Poincaré sections as a function of wave magnetic field amplitude. For simplicity, we assume a uniform background magnetic field. We specify $k_\perp \rho_i = 3$, $\omega = \Omega_i/5$, $\beta = 1$, $T_e/T_i = 0.2$ and $k_\parallel \rho_i \approx 0.05$ consistent with the kinetic Alfvén wave dispersion relation. The small perpendicular scales are consistent with typical kinetic Alfvén wave solutions at the magnetopause with wavelength the order of 100km $\approx 2\rho_i$ (*Johnson and Cheng*, 1997b; *Johnson et al.*, 2001).

For clarity of Poincaré section plots, we take the wave frequency equal to 1/5 of the ion gyrofrequency. For smaller frequency similar physics leads to stochastic thresholds, but the formation of island chains composed of hundreds of islands cannot be as easily seen by eye. Interestingly, the stochastic threshold primarily depends on wave amplitude and is not strongly dependent on wave frequency in the range of interest. Note that because $k_\perp \rho_i$ is larger than 1, the kinetic Alfvén wave does have a significant electrostatic component associated with ion Larmor radius effects.

The Poincaré section for $\delta B_\perp/B_0 = 0.006$ is shown in Figure 5. Particles are started with $\Psi = 0$ with varying initial value of μ and $v_\parallel = 0$. If there were no wave, the particles would simply gyrate with constant value of μ. For small wave amplitude, $\mathbf{X}$ is nearly periodic in the gyrophase, so dependence of Ψ on time is primarily through $-\omega t$. Because the wave frequency is $\Omega_i/5$, Ψ will approximately decrease by $2\pi/5$ each gyroperiod until it returns to the original phase (minus 2π). The Poincare section for a given particle would therefore reduce to five equally spaced points at constant μ. However, with the addition of the wave, the particle gyration can be retarded or accelerated.

In Figure 5, it is apparent that the for initial $\mu < 0.41$ particles do not return to the same wave phase after a wave period (five gyroperiods), but have a small positive increment in phase Ψ. On the other hand, for initial $\mu > 0.41$, it is apparent that the after a wave period,

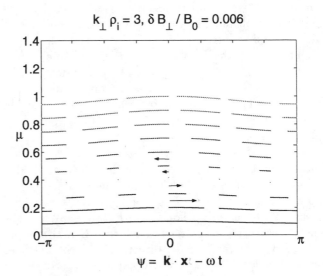

$$k_\perp \rho_i = 3, \ \delta B_\perp / B_0 = 0.006$$

Figure 5. Poincaré section for small amplitude wave. Note that low energy particles are advanced in the phase of the wave while higher energy particles are retarded. Phase space islands with period five first emerge along the stationary phase trajectory.

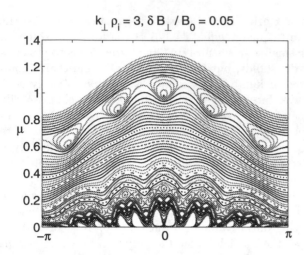

$k_\perp \rho_i = 3, \delta B_\perp / B_0 = 0.05$

Figure 6. As the wave amplitude is increased, chains of phase space islands appear. The period 16 island chain has just begun to merge and a separatrix appears.

the phase, Ψ, of the particle will have a small negative increment. Near initial $\mu = 0.41$, it is evident that there is a boundary where the orbit is stationary and the Poincaré section only consists of only five points. As the amplitude is increased these fixed points move to larger values of μ and nearby orbits circulate around the fixed points. Moreover, other boundaries appear across which the phase advances/retards, but with higher order periodicity. This phenomenon is illustrated in Figure 6 which shows the Poincaré section for $\delta B_\perp / B_0 = 0.05$. The period five island chain associated with the original transition boundary has moved to larger value of μ and is quite large in extent. Above that chain, the high energy particles do not show any structure related to the gyromotion and simply float up and down in the wave. Other islands chains have also appeared at lower energies. The obvious island periods evidently are in the sequence: 26, 21, 58, 16, 27 and so forth. Islands in the period 16 and 21 chains have just begun to overlap.

At the lowest energies, a chain of five fingers (rather than islands) has appeared. The fingers have divided into ten subfingers. The central finger contains two islands.. The physical origin of the fingers is the increase of the $\mathbf{E} \times \mathbf{B}$ velocity of the wave. When low energy particles are subjected to large $\mathbf{E} \times \mathbf{B}$ motion, the gyrophase can reverse direction and the particles are trapped in the wave. As a result, the particles are not sampled during their gyromotion and the island is incomplete (that is, the low energy particles can skip one or more fingers during their gyromotion).

A slight increase in wave amplitude shown in Figure 7 allows the phase space islands to merge and re-

gions of stochastic orbits appear. Island structures still remain embedded in the stochastic region, but now particle trajectories can wander through the stochastic sea to higher energies than previously accessible. Hence, the low energy part of the ion distribution can be effectively heated. Note that the ten finger structures have now moved into the stochastic sea and comprise a sequence of ten islands. A new five fingered structure has also begun to emerge from the low energy part of the phase space with regular orbits below the stochastic sea with a fixed point in the central island. The stationary orbit at $\Psi = 0$ in the central finger is an elongated orbit which has period equal to that of the wave and can be considered to be in nonlinear resonance with the wave. Above the stochastic sea, particle trajectories lie on well defined curves or island chains and the particles are not heated. The large period five island chain remains intact and moves to higher energy. However, clear boundaries still confine heating to the lowest energy ions.

For $\delta B_\perp / B_0 \sim 0.11$ as shown in Figure 8 nearly the entire low μ region becomes stochastic except for a few small islands that remain embedded in the stochastic sea. The stochastic region is forced up against the set of period five islands which have themselves become stochastic. A clear boundary exists between the two stochastic regions and particles may not move across that boundary. Above the boundary, the period five is-

$k_\perp \rho_i = 3, \delta B_\perp / B_0 = 0.056$

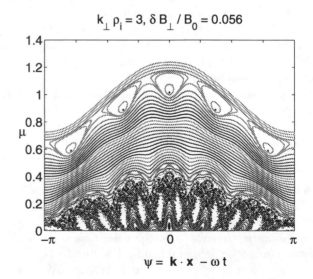

$\psi = \mathbf{k} \cdot \mathbf{x} - \omega t$

Figure 7. With a slight increase in wave amplitude, the low energy region of the Poincaré section becomes stochastic. Low energy particles can now wander through the stochastic sea to higher energies. Phase space islands remain embedded in the stochastic sea, but the entire region becomes stochastic for $\delta B / B_0 = 0.056$.

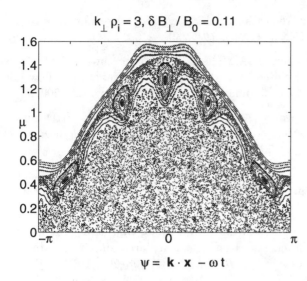

$k_\perp\,\rho_i = 3,\ \delta\,B_\perp\,/\,B_0 = 0.11$

$\psi = \mathbf{k}\cdot\mathbf{x} - \omega\,t$

Figure 8. Both the period five island chain and the low energy region are stochastic, but they are separated by a boundary. Global stochasticity occurs when the two regions merge just above $\delta B_\perp/B_0 = 0.11$.

land chain has also merged and become stochastic and other chains of islands have also appeared against the boundary between the lower stochastic region and the upper stochastic region. Chains of islands have also appeared inside the period five islands. However, the two regions are still separated and low energy particles cannot be energized much above the thermal speed. With a modest increase to $\delta B_\perp/B_0 = 0.14$ as shown in Figure 9, the period five islands merge with the low energy stochastic region leaving a path for low energy particles to be heated well beyond the thermal energy. With further increase in the wave amplitude the stochastic regime pushes to larger μ. For example, with $\delta B_\perp/B_0 \sim 0.3$ particles are readily energized to $\mu \sim 4$.

These results have several implications relevant to magnetopause observations. First, transverse ion heating due to this process depends on wave amplitude. Below the wave threshold, there is no heating of the plasma. Slightly above the threshold, the core of the distribution is expected to flatten, but ions are not heated above the thermal velocity. For larger wave amplitude, ions can be heated above the thermal velocity. The heating process can occur rapidly over a time less than 30 cyclotron periods. Due to the brevity of this letter, we defer estimates of the stochastic threshold, more detailed description of the appearance of island chains, island overlap,, and discussion of the effect of background gradients (and the resulting particle transport) for a later publication.

Obviously, wave heating at the magnetopause is more complicated than this simple picture. This calculation is primarily meant to provide understanding of the non-linear coupling between kinetic Alfvén waves and cyclotron motion and to provide a physical picture which gives qualitative understanding of resulting particle signatures. Most likely, there is a spectrum of waves that can participate in the heating process. Typically, the addition of a second wave or magnetic field rotation can reduce the threshold for stochasticity to occur and increase particle heating and transport beyond quasi-linear levels (*Johnson and Cheng*, 1997c).

6. SUMMARY

In this paper we have examined the role of kinetic Alfvén waves at the magnetopause. ULF waves dominate the spectrum of nearly every magnetopause crossing. Probably the most distinctive wave features at the magnetopause are the large increase in wave amplitude and abrupt change in wave polarization where the magnetopause Alfvén velocity gradient is encountered. The abrupt change in wave characteristics is suggestive of a process which could convert the compressional wave into a shear wave. Theoretical examination of the mode conversion process yields results similar to the observations. Compressional wave mode convert into kinetic Alfvén waves at the magnetopause. The mode structure suggests that wave amplitude will be increased and polarization will change abruptly. Magnetic ro-

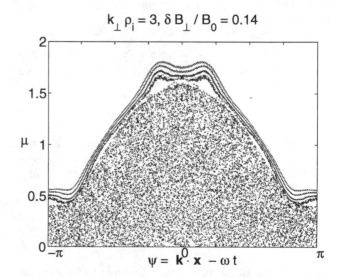

$k_\perp\,\rho_i = 3,\ \delta\,B_\perp\,/\,B_0 = 0.14$

$\psi = \mathbf{k}\cdot\mathbf{x} - \omega\,t$

Figure 9. Global stochasticity for $\delta B_\perp/B_0 = 0.14$. Cold and/or thermal particles may wander to superthermal energies resulting in heating perpendicular to the magnetic field.

tation across the magnetopause provides a good test of the theory of resonant mode conversion. Wave observations appear to be in rough agreement with the theory which predicts small wave amplification below a threshold shear angle and roughly constant amplification above the shear angle. The amplification as a function of frequency also seems to agree qualitatively with observations. It therefore seems reasonable to believe that the ULF wave activity at the magnetopause is the result of a mode conversion process.

That a population of large amplitude ULF waves nearly always populates the magnetopause and the associated boundary layers is not without consequence. Quasilinear theory predicts substantial plasma transport for typical observed parameters. Moreover, because the waves have such large amplitude, the orbits can also be stochastic leading to rapid transport and heating. We examined single particle orbits in kinetic Alfvén waves and demonstrated that ions can be significantly heated for modest wave amplitudes. Recent particle observations near the magnetopause and in the low-latitude boundary layer show significant heating of low energy ions without significant heating of the energetic population. Such observations are a typical signature of stochastic ion heating near the stochastic threshold. Moreover, in plasma with magnetic shear, stochastic ion transport can even exceed the quasilinear particle entry estimate of 10^{27} particles/sec. Hence, kinetic Alfvén waves found at the magnetopause contribute to the population of the boundary layers and can explain some of the observed wave and particle signatures found in those layers.

Acknowledgments.   This work is supported by the NSF grant ATM-9906142 and DoE Contract CGLNo. DE-AC02-76-CHO3073.

REFERENCES

Anderson, B. J., S. A. Fuselier, and D. Murr, Electromagnetic ion cyclotron waves observed in the plasma depletion layer, *Geophys. Res. Lett.*, *18*, 1955, 1991.

Belmont, G., F. Reberac, and L. Rezeau, Resonant amplification of magnetosheath MHD fluctuations at the magnetopause, *Geophys. Res. Lett.*, *22*, 295–298, 1995.

Cheng, C. Z., and J. R. Johnson, A kinetic-fluid model, *J. Geophys. Res.*, *104*, 413–427, 1999.

De Keyser, J., M. Roth, F. Reberac, L. Rezeau, and G. Belmont, Resonant amplification of MHD waves in realistic subsolar magnetopause configurations, *J. Geophys. Res.*, *104*, 2399–2409, 1999.

Fuselier, S. A., The LLBL for northward IMF: Open or closed, AGU Chapman Conference, The Low Latitude Boundary Layer, New Orleans, LA, 2001.

Hasegawa, A., Particle acceleration by MHD surface wave and formation of aurora, *J. Geophys. Res.*, *81*, 5083–5090, 1976.

Hasegawa, A., and L. Chen, Parametric decay of kinetic Alfvén wave and its application to plasma heating, *Phys. Rev. Lett.*, *36*, 1362–1365, 1976.

Hsu, J. Y., K. Matsuda, M. S. Chu, and T. H. Jensen, Stochastic heating of a large-amplitude standing wave, *Phys. Rev. Lett.*, *43*, 203–206, 1979.

Johnson, J. R., and C. Z. Cheng, Global structure of mirror modes in the magnetosheath, *J. Geophys. Res.*, *102*, 7179–7189, 1997a.

Johnson, J. R., and C. Z. Cheng, Kinetic Alfvén waves and plasma transport at the magnetopause, *Geophys. Res. Lett.*, *24*, 1423–1426, 1997b.

Johnson, J. R., and C. Z. Cheng, Plasma transport at the magnetopause due to low frequency mhd waves in a strongly sheared magnetic field, *Eos Trans. AGU*, *78*(46), F591, Fall Meet. Suppl., 1997c.

Johnson, J. R., and C. Z. Cheng, Signatures of mode conversion and kinetic Alfvén waves at the magnetopause, *Geophys. Res. Lett.*, *28*, 227–230, 2001.

Johnson, J. R., C. Z. Cheng, and P. Song, Signatures of mode conversion and kinetic Alfvén waves at the magnetopause, *Geophys. Res. Lett.*, *28*(2), 227–230, 2001.

Karney, C. F. F., and A. Bers, Stochastic ion heating by a perpendicularly propagating electrostatic wave, *Phys. Rev. Lett.*, *39*, 550, 1977.

Lee, L. C., J. R. Johnson, and Z. W. Ma, Kinetic Alfvén waves as a source of plasma transport at the dayside magnetopause, *J. Geophys. Res.*, *99*, 17405–17411, 1994.

Perraut, S., R. Gendrin, P. Robert, and A. Roux, Magnetic pulsations observed onboard GEOS 2 in the ULF ran ge during multiple magnetopause crossings, in *Eur. Space Agency Spec. Publ.*, *148*, pp. 113–122. 1979.

Phan, T., Formation of the LLBL by merging, AGU

Chapman Conference, The Low Latitude Boundary Layer, New Orleans, LA, 2001.

Phan, T. D., and G. Paschmann, Low-latitude dayside magnetopause and boundary layer for high magnetic shear 1. structure and motion, *J. Geophys. Res.*, *101*, 1996.

Phan, T. D., et al., Low-latitude dusk flank magnetosheath, magnetopause, and boundary layer for low magnetic shear: Wind observations, *J. Geophys. Res.*, *102*, 1997.

Rezeau, L., A. Roux, and C. T. Russell, Characterization of small-scale structures at the magnetopause from ISEE measurements, *J. Geophys. Res.*, *98*, 179, 1993.

Song, P., Observations of waves at the dayside magnetopause, in *Solar Wind Source of Magnetospheric Ultra-Low-Frequency Waves, Geophysical Monograph Series*, vol. 81, pp. 159–171. 1994.

Song, P., et al., Wave properties near the subsolar magnetopause for northward interplanetary magnetic field: Multiple instrument particle observations, *J. Geophys. Res.*, *98*, 11,319–11,337, 1993a.

Song, P., C. T. Russell, and C. Y. Huang, Wave properties near the subsolar magnetopause: Pc 1 waves in the sheath transition layer, *J. Geophys. Res.*, *98*, 5907–5923, 1993b.

Song, P., C. T. Russell, R. J. Strangeway, J. R. Wygant, C. A. Cattell, R. J. Fitzenreiter, and R. R. Anderson, Wave properties near the subsolar magnetopause: Pc 3–4 energy coupling for northward interplanetary magnetic field, *J. Geophys. Res.*, *98*, 187–196, 1993c.

Wilber, M., G. K. Parks, and R. P. Lin, Conic-like ion components observed interior to the dusk magnetopause, AGU Chapman Conference, The Low Latitude Boundary Layer, New Orleans, LA, 2001.

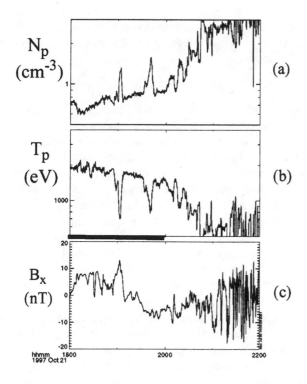

Figure 6. Zoom-in of the Geotail observations in the 18:00 - 22:00 UT interval on October 21, 1997. (a) Ion density; (b) ion temperature; (c) the x component of the magnetic field.

IMF, *Terasawa et al.* [1997] suggested that a diffusion-type process is working at the flank low-latitude boundary layer (LLBL). The location of the cold dense plasma sheet along the flanks, next to the magnetopause (Plate 2), is suggestive of an LLBL source for this plasma. In addition, the low average flow speed in the cold dense plasma sheet and its large standard deviation (Plate 3) is an expected signature of (turbulent) diffusive entry, although the expected inward average motion of the plasma sheet plasma is not seen in our study (section 3.4). If the diffusive entry becomes more effective with downtail distance it may explain the observed confinement of cold dense plasma sheet to high latitudes (Figure 2). However, present theoretical models cannot account for LLBL thicknesses greater than ~ 0.5 R_E [e.g., *Treumann et al.*, 1995], while the present observations imply that the cold dense plasma sheet region is substantially thicker and can extend inward to $|Y_{GSM}| \sim 7$ R_E (Plate 2).

Fujimoto and Terasawa [1994], and later *Fairfield et al.* [2000] and *Otto and Fairfield* [2000], suggested that Kelvin-Helmholtz instabilities on the flank magnetopause could add cold dense plasma to closed magnetospheric field lines. Such a scenario cannot be verified with the present observations since multi-spacecraft observations and/or tomography imaging are needed to reveal the possible presence of Kelvin-Helmholtz vortices along the flank magnetopause.

5.2. Magnetic Reconnection

Standard low-latitude dayside magnetic reconnection result in solar wind plasma entry confined to the the reconnection layer itself, which is typically a fraction of an Earth radius [*Paschmann et al.*, 1986; *Gosling et al.*, 1986; *Phan and Paschmann*, 1996]. However, other reconnection scenarios have been proposed which could give rise to substantially larger solar wind plasma entry across the dayside and flank magnetopause. Here we discuss the tail-flank boundary layer model by *Raeder et al.* [1995; 1997] based on the qualitative model by *Song and Russell* [1992], and the "magnetospheric sash" model by *White et al.* [1998] and *Nishikawa* [1998].

5.2.1. Tail-flank boundary layer model for northward IMF. Global MHD simulations by *Raeder et al.* [1995; 1997] revealed the formation of a broad boundary layer (which they named the tail flank boundary layer) that consists of closed magnetic field regions formed by reconnection between IMF and lobe field lines, confirming the qualitative model by *Song and Russell* [1992]. According to this model, the northward-directed IMF reconnects with lobe field lines in the high-latitude magnetopause tailward of the cusps in both hemispheres. Closed magnetic flux tubes are formed out of field lines that were originally in the solar wind, one end of the field lines connecting to the northern hemisphere and the other end to the southern hemisphere. The newly created closed flux tube is convected tailward via the LLBL, producing a growing LLBL that protrudes more and more into the magnetotail volume. Thus this model describes the "capturing" by reconnection of solar wind plasma to form an extended tail flank tailward-flowing boundary layer on closed field lines that could fill a substantial part of the plasma sheet.

The observations of cold dense plasma sheet regions extending several R_E inward from the flanks is consistent with the *Raeder et al.* [1995; 1997] model (Plate 2). However,

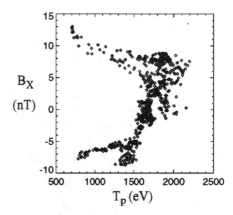

Figure 7. B_x component versus ion temperature observed by Geotail in the 18:00-20:00 UT interval (marked by the thick horizontal line in Figure 6)

the Raeder et al. model does not predict the observed confinement of cold dense plasma to high latitudes (Figure 2) nor does it predict the plasma to be stagnant (Plate 3). The tail flank boundary layer is predicted to move anti-sunward [*Raeder et al.*, 1997].

5.2.2. The "sash". In the global MHD models of *White et al.* [1998] and *Nishikawa* [1998], a "sash" is created along a band of weak magnetic field representing the locus of points along the last open field lines. The sash represents the sites where direct access of magnetosheath plasma across the magnetopause can occur. In these models reconnection occur from the cusp through the tail flanks, facilitating direct entry of magnetosheath particles into the plasma sheet [*Nishikawa*, 1998]. The location of the sash depends strongly on the IMF orientation. For a strong duskward IMF, the sash would run tailward along the high-latitude magnetopause dusk flank from the northern cusp to the southern cusp on the dawnside, closing via the cross-tail neutral sheet. For a dawnward-directed IMF, the situation is reversed in terms of the dawn-dusk and north-south locations of the sash.

For the Wind data set the spatial occurrence of the cold dense plasma sheet does not depend on the IMF B_y component (not shown). Cold dense plasma is observed both at the duskside and at the dawnside, in both the northern and southern hemisphere for both IMF B_y directions. However, a lack of IMF B_y dependence does not exclude the possibility that the cold dense plasma sheet originates on open field lines and then migrates onto closed field lines (by yet unknown mechanisms), causing cold dense plasma that originally entered from one hemisphere to eventually occur in both hemispheres, masking the initial IMF B_y dependence.

5.3. Electric Fringe Field

Lennartsson [1992] suggested that solar wind plasma can enter along the tail flanks in a region between the tail lobes and the plasma sheet during northward IMF, and this plasma is then convected inward by the electric fringe field in the LLBL. This model predicts a significant velocity component directed away from the nearest tail flank. The observations of cold dense plasma sheet only at high latitudes is in agreement with the *Lennartsson* [1992] model. However, the low flow speeds in the y direction observed in the cold dense plasma sheet (Plate 3b and 5c) do not appear to be consistent with the electric fringe field effect, which is expected to produce significant inward plasma flows from the flanks.

5.4. Magnetic Gradient/Curvature Drift

Spence and Kivelson [1993] developed a convection model for the magnetotail including the dawnside LLBL as one of the plasma sources, taking into account the duskward gradient/curvature drift. However, the fact that the cold dense plasma sheet is observed on both the dawnside and the duskside (Plate 2) is inconsistent with the *Spence and Kivelson* [1993] model.

5.5. Summary of Model Comparison

A tail flank entry process is likely to be significant in populating the plasma sheet during northward IMF and quiet geomagnetic conditions, and should be less important during active times when reconnection in the distant tail is expected to be the dominant solar wind entry process [*Dungey*, 1961; *Cowley*, 1980]. Most of the models of plasma entry through the tail flank magnetopause discussed above require further testing and cannot be verified using the present observations. However, one model, the magnetic gradient/curvature drift model [*Spence and Kivelson*, 1993], cannot explain the presence of cold dense plasma sheet on both the dawn flank and the dusk flank, since it predicts solar wind entry through the dawnside LLBL only.

Other models appear to be able to explain some, but not all, of the observed features. Present theoretical diffusion models cannot account for LLBL thicknesses greater than ~0.5 R_E [e.g., *Treumann et al.*, 1995], but if the diffusive entry becomes more effective with downtail distance it may explain the confinement of cold dense plasma sheet to high latitudes. The large standard deviation for the cold dense plasma sheet velocity (Plate 3) may also indicate turbulent mixing. The entry of solar wind plasma by the capture of Kelvin-Helmholtz vortices [*Fujimoto and Terasawa*, 1994] cannot be verified without multi-spacecraft observations or tomography imaging.

Simulations have shown that magnetic reconnection poleward of the cusp captures magnetosheath plasma onto closed magnetospheric field lines that are convected tailward, creating a thick tail flank boundary layer [*Raeder et al.*, 1995; 1997]. However, this tail flank boundary layer is predicted to be tailward-flowing and not confined to high latitudes, and cannot fully explain all observed features. The electric fringe field model where solar wind plasma can enter the plasma sheet between the tail lobes and the plasma sheet during northward IMF [*Lennartsson*, 1992] is consistent with the observations of cold dense plasma sheet predominantly on high latitudes. However, this model cannot explain the fact that the cold dense plasma sheet is nearly stagnant, with little or no inward (away from the flanks) flow. As for the sash model [*White et al.*, 1998]; *Nishikawa*, 1998], we do not find any direct evidence for the sash as a source of the cold dense plasma sheet.

Future studies involving multi-spacecraft and statistical studies, as well as tomography imaging, are expected to shed more light on the solar wind entry processes during northward IMF. In the current study we have not made a distinction between single and double population (mixed) cold dense ion distributions, as those studied by e.g., *Fujimoto et al.* [1996; 1998], *Fuselier et al.* [1999], and *Phan et al.* [2000]. A survey of the occurrence and properties of the mixed ions will be a topic for an upcoming study.

Acknowledgments. We would like to thank T. Mukai and S. Kokubun, the principal investigators of the Geotail particle and magnetic field experiment, respectively. We are grateful to Ronald P. Lepping for providing Wind and IMP-8 magnetic field data. Solar wind magnetic field data were obtained from the ACE science center website. We thank the ACE MAG experiment team for making their data available. This research was funded in part by NASA grants NAG5-10471 and NAG5-6928 at UC Berkeley.

REFERENCES

Baumjohann, W., et al., Average ion moments in the plasma sheet boundary layer, *J. Geophys. Res., 93*, 11,507, 1988.

Baumjohann, W., et al., Average plasma properties in the central plasma sheet, *J. Geophys. Res., 94*, 6597, 1989.

Borovsky, J. E., M. F. Thomsen, and R. C. Elphic, The driving of the plasma sheet by the solar wind, *J. Geophys. Res., 103*, 17617, 1998.

Cowley, S. W. H., Plasma populations in a simple open model magnetosphere, *Space Sci. Rev.*, 26, 217, 1980.

Dungey, J. W., Interplanetary magnetic field and the auroral zones, *Phys. Rev. Lett., 6*, 47, 1961.

Eastman, T. E., L. A. Frank, and C. Y. Huang, The boundary layers as the primary transport regions of the earth's magnetotail, *J. Geophys. Res., 90*, 9541, 1985.

Fairfield, D. H., et al., Simultaneous measurements of magnetotail dynamics by IMP spacecraft, *J. Geophys. Res., 86*, 1396, 1981.

Fairfield, D. H., et al., Geotail observations of the Kelvin-Helmholtz instability at the equatorial magnetotail boundary for parallel northward fields, *J. Geophys. Res., 105*, 21,159, 2000.

Fujimoto, M., and T. Terasawa, Anomalous ion mixing within an MHD scale Kelvin-Helmholtz vortex, *J. Geophys. Res., 99*, 8601, 1994.

Fujimoto, M., et al., Plasma entry from the flanks of the near-Earth magnetotail: Geotail observations in the dawnside LLBL and the plasma sheet, *J. Geomagn. Geoelectr., 48*, 711, 1996.

Fujimoto, M., et al., Plasma entry from the flanks of the near-Earth magnetotail: Geotail observations, *J. Geophys. Res., 103*, 4391, 1998.

Fuselier, S. A., R. C. Elphic, and J. T. Gosling, Composition measurements in the dusk flank magnetosphere, *J. Geophys. Res., 104*, 4515, 1999.

Gosling, J. T., et al., Accelerated plasma flows at the near-tail magnetopause. *J. Geophys. Res., 89*, 6689, 1986.

Kokubun, S., et al., The Geotail magnetic field experiment, *J. Geomagn. Geoelectr., 46*, 7, 1994.

Lennartsson, W., A scenario for solar wind penetration of earth's magnetic tail based on ion composition data from the ISEE 1 spacecraft, *J. Geophys. Res., 97*, 19,221, 1992.

Lennartsson, W., Statistical investigation of IMF B_z effects on energetic (0.1- to 16-keV) magnetospheric O + ions, *J. Geophys. Res., 100*, 23,621, 1995.

Lennartsson, O. W., and E. G. Shelley, Survey of 0.1- to 16-keV/e plasma sheet ion composition, *J. Geophys. Res., 91*, 3061, 1986.

Lepping, R. P., et al., The Wind magnetic field investigation, *Space Sci. Rev., 71*, 207, 1995.

Lin, R. P., et al. A three-dimensional plasma and energetic particle investigation for the Wind spacecraft, *Space Sci. Rev., 71*, 125, 1995.

McComas, D. J., et al., Solar Wind Electron Proton Alpha Monitor (SWEPAM) for the Advanced Composition Explorer, *Space Sci. Rev., 86*, 563, 1998.

Mukai, S. et al., The low energy particle (LEP) experiment onboard the GEOTAIL satellite, *J. Geomagn. Geoelectr., 46*, 669, 1994.

Nishikawa, K. I., Particle entry through reconnection grooves in the magnetopause with a dawnward IMF as simulated by a 3-D EM particle code, *Geophys. Res. Lett.*, 25, 1609, 1998.

Øieroset, M., et al., Energetic ion outflow from the dayside ionosphere: Categorization, classification, and statistical study, *J. Geophys. Res., 104*, 24,915, 1999.

Otto, A., and D. Fairfield, Kelvin-Helmholtz instability at the magnetotail boundary: MHD simulation and comparison with Geotail observations, *J. Geophys. Res., 105*, 21,175, 2000.

Paschmann, G., et al., The magnetopause for large magnetic shear - AMPTE/IRM observations, *J. Geophys. Res., 91*, 11,099, 1986.

Paterson, W. R., and L. A. Frank, Survey of plasma parameters in the Earth's distant magnetotail with the Geotail spacecraft, *Geophys. Res. Lett., 21*, 2971, 1994..

Phan, T. D., and G. Paschmann, Low-latitude dayside magnetopause and boundary layer for high magnetic shear 1. Structure and motion, *J. Geophys. Res., 101*, 7801, 1996.

Phan, T. D., et al., Wind observations of the halo/cold plasma sheet, in *Substorms-4*, edited by S. Kokubun and Y. Kamide, Terra Scientific/Kluwer Academic, Boston, 1998.

Phan, T. D., et al., Wind observations of mixed magnetosheath-plasma sheet ions deep inside the magnetosphere, *J. Geophys. Res., 105*, 5497, 2000.

Raeder, J., R. J. Walker, and M. Ashour-Abdalla, The structure of the distant geomagnetic tail during long periods of northward IMF, *Geophys. Res. Lett., 22*, 349, 1995.

Raeder et al., Boundary layer formation in the magnetotail: Geotail observations and comparisons with a global MHD simulation, *Geophys. Res. Lett., 24*, 951, 1997.

Rosenbauer, H., et al., Heos 2 plasma observations in the distant polar magnetosphere: The plasma mantle, *J. Geophys. Res., 80*, 2723, 1975.

Sckopke, N., et al., Structure of the low-latitude boundary layer, *J. Geophys. Res., 86*, 2099, 1976.

Sckopke, N., et al., Structure of the low-latitude boundary layer, *J. Geophys. Res., 86*, 2099, 1981.

Smith, C. W., et al., The ACE Magnetic Fields Experiment, *Space Sci. Rev., 86*, 613, 1998.

Song, P., and C. T. Russell, A model of the formation of the low-latitude boundary layer, *J. Geophys. Res., 94*, 1411, 1992.

Spence, H. E., and M. G. Kivelson, Contributions of the low-latitude boundary layer to the finite width magnetotail convection model, *J. Geophys. Res., 98*, 15487, 1993.

Terasawa, T., et al., Solar wind control of density and temperature in the near-Earth plasma sheet: WIND/GEOTAIL collaboration, *Geophys. Res. Lett., 24*, 935, 1996.

Treumann, R. A., J. Labelle, and T. M. Bauer, Diffusion at the magnetopause: The observational viewpoint, in *Physics of the Magnetopause, Geophys. Monogr. 90*, edited by P. Song, B. U. Ö. Sonnerup, and M. F. Thomsen, AGU, Washington, D. C., p. 331, 1995.

White, W. W., et al., The magnetospheric sash and the cross-tail S, *Geophys. Res. Lett., 25*, 1605, 1998.

Wing, S., and P. T. Newell, Central plasma sheet ion properties as inferred from ionospheric observations, *J. Geophys. Res., 103*, 6785, 1998.

Yau, A., Energetic auroral and polar ion outflow at DE-1 altitudes: Magnitude, composition, magnetic activity dependence and long-term variations, *J. Geophys. Res., 90*, 8417, 1985.

Zwickl, R. D., et al., Evolution of the Earth's distant magnetotail, ISEE-3 electron plasma results, *J. Geophys. Res., 89*, 11,007, 1984.

L. Chan, R. P. Lin, M. Øieroset, and T. D. Phan, Space Sciences Laboratory, University of California, Berkeley, CA, 94720. (e-mail: oieroset@ssl.berkeley.edu)

M. Fujimoto, Department of Earth and Planetary Sciences, Tokyo Institute of Technology, Meguro, Tokyo 152-8551, Japan. (e-mail: fujimoto@geo.titech.ac.jp)

R. Skoug, Group NIS-1 MS D466, Los Alamos National Laboratory, Los Alamos, NM 87545. (e-mail: rskoug@lanl.gov)

Observational Signatures of Plasma Transport across the Low-latitude Boundary Layer

Masaki N. Nishino, Toshio Terasawa

Department of Earth and Planetary Science, University of Tokyo, Tokyo, Japan

Masaki Fujimoto

Department of Earth and Planetary Sciences, Tokyo Institute of Technology, Tokyo, Japan

It has been recognized that during extended periods of the northward interplanetary magnetic field the tail plasma sheet becomes cold and dense, showing a positive density correlation with the solar wind plasma. Recently it has been also recognized that the plasma density integrated along the Z (north-south) direction across the plasma sheet becomes also high during the northward IMF periods, which suggests a fairly high plasma supply rate of $\sim 10^{26}$ protons/sec amounting nearly 10% of the enhanced supply rate during the southward interplanetary magnetic field periods. While the latter rate is considered to be caused by the efficient dayside magnetopause reconnection, it is not yet known how the plasma transport occurs during the northward interplanetary magnetic field periods. Since the highly evolved LLBL is also observed during such periods, we expect some causal relation between the plasma transport to the plasma sheet and the evolution of the LLBL. We review the key observations and discuss possible physical mechanisms of the plasma transportation.

1. INTRODUCTION

Plasma transport processes across the magnetopause are thought to be closely connected to the formation of the LLBL. Two physical mechanisms have been proposed to explain the formation of the LLBL; One is diffusion-type process at magnetotail flanks, and the other is magnetic reconnection which is expected to occur at high-latitude magnetopause during the northward interplanetary magnetic field (N-IMF, hereafter) intervals. However, what is really going on at the tail LLBL under the N-IMF condition has still been an open question.

The idea that the flank region plays the primary role in transporting the solar wind plasma into the plasma sheet has been proposed by several authors [Eastman et al.,1985; Lennartsson,1992; Fujimoto et al.,1996]. These authors expected the occurrence of diffusion-type process at the magnetopause. Recently, several authors have presented evidence of the anomalous magnetotail behavior during strongly northward IMF periods [Chen et al.,1993; Fairfield,1993; Fairfield et al.,1996; Fujimoto et al.,1998]. Their observations suggest that the closed field region between the plasma sheet proper and the magnetosheath becomes quite turbulent under such N-IMF condition.

High-latitude magnetic reconnection is another candidate which is expected to produce the structure of the LLBL [Crooker, 1979; Song and Russell, 1992]. In this process, solar wind and magnetospheric field lines reconnect in the high latitude magnetopause at both hemispheres. Closed loops of magnetic flux are formed

Earth's Low-Latitude Boundary Layer
Geophysical Monograph 133
Copyright 2003 by the American Geophysical Union
10.1029/133GM26

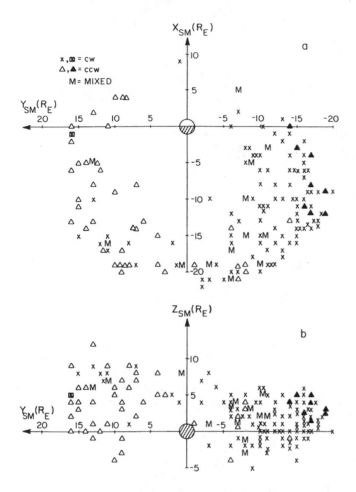

Figure 1. Distribution of occurrences of vortical plasma flow [Hones et al., 1981]. Instances of clockwise (CW) and counterclockwise (CCW) flow are indicated by X's and triangles, respectively.

out of solar wind flux, one end of the loop connecting to the northern and one to the southern hemispheres. The closed flux is transported tailward via the LLBL. The solar wind thus contributes to the plasma density enhancement observed in the tail by producing a growing LLBL that protrude more and more into the magnetotail volume.

It is widely accepted that during the southward IMF periods magnetic reconnection plays an important role in plasma transport from the solar wind into the magnetosphere and in turn in the magnetotail plasma sheet with an enhanced plasma convection flow. On the other hand, such a dynamic plasma motion in the plasma sheet does not exist during the N-IMF intervals. In past studies it has been expected that at or near the magnetopause boundary layer Kelvin-Helmholtz instability occurs during the N-IMF intervals and can play a role in plasma transport from the solar wind into the LLBL. From the viewpoint of Kelvin-Helmholtz insta-

bility Hones et al. [1981] investigated the rotations of the plasma bulk flow in the night side plasma sheet and found that the preferred sense of the rotations was clockwise (CW) as viewed from above the dawnside and counterclockwise (CCW) on the duskside (Figure 1). They concluded that the rotations could be explained by "vortex structures" in the XY plane, which were believed to be produced by the Kelvin-Helmholtz instability due to the sheared plasma flow between the plasma sheet and the boundary layer. Recently, a duskside magnetopause crossing event with clear rotations of plasma bulk flow has been reported [Fairfield et al., 2000; Nishino et al., 2001a]. These observations imply that rotational motions at or near the matnetopause boundary layer can be caused by Kelvin-Helmholtz instability and accompanied vortical structure can transport plasma particles from the magnetosheath into the LLBL, and that "vortex structures" deep inside the plasma sheet also play an important role in plasma transport in the magnetotail.

Furthermore, it has been known that cold dense plasma (CDP; $N_i > 1$/cc, $T_i < 1$ keV) is sometimes observed in the magnetotail plasma sheet during the N-IMF periods [Fujimoto et al., 1998]. These observations of cold dense plasma sheet (CDPS) is thought to be closely related to plasma transport during the N-IMF intervals. Since the CDPS is observed not only near the flank region but also deep inside the plasma sheet, we first review recent statistical analysis of the cold dense plasma sheet, and next case studies which well explain the status of the LLBL and the plasma sheet under very northward IMF condition.

2. STATISTICAL STUDIES OF THE DENSE PLASMA SHEET

Terasawa et al. [1997] quantitatively confirmed that there is a positive correlation between the north-south angle of the IMF and the proton density normalized by the solar wind density in the near-earth and middle-distant plasma sheet ($-15 > X > -50$ R_E). The best correlations between the plasma sheet and solar wind parameters during the N-IMF periods occur when the latter quantities are averaged over 9^{+3}_{-4} hours prior to the plasma sheet observations (Figure 2). Near the dawn and dusk flanks of the plasma sheet temperature decreases and density increases. Since these effects were found to be more significant in the flank region of the tail, it is suggested that during prolonged N-IMF periods ($\sim$ several hours) there is a slow diffusive transport of the plasma from the solar wind into the plasma sheet through the magnetotail flanks.

Another example is 'superdense plasma sheet' (SDPS) found in the inner magnetosphere by Borovsky et al. [1997]. They have reported that a few days per month the plasma sheet at the geosynchronous orbit has a

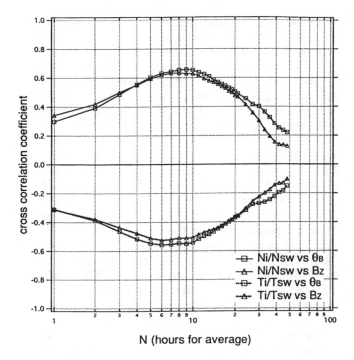

Figure 2. Cross correlations (solid [dotted] curves) between the normalized plasma sheet density [temperature] and N-hour-averaged IMF parameters, θ_B^N and $B_{SW,z}^N$ [Terasawa et al., 1997].

fairly high plasma density ~ 3 particles cm^{-3}, which is 10 times larger than the average. It is noteworthy that SDPS observations are in conjunction with sharp rises in the Kp index after it has been low for an extended period (Figure 3). Borovsky et al. [1997] suggested two candidate processes: (1) Plasmaspheric flux tubes are transported from the dayside over the poles and into the tail during the early phases of magnetic storms due to dayside reconnection (Elphic et al., 1997; Freeman et al., 1977). (2) The high density solar wind enters directly into the plasma sheet.

Borovsky et al. [1997] referred to 1.5-hour-long pulse of high-density solar wind during the CDAW 6 event of 22 March 1979. They attributed the occurrence of a brief superdense plasma sheet to the high-density solar wind pulse. They argued that composition measurements of this high-density pulse in the magnetotail and at gyosynchronous orbit [Baker et al. 1985; Lennartsson et al., 1985; Stockholm et al., 1985; Ipavich et al., 1985] indicate that it is probably of solar-wind origin.

Later in this review, we will discuss a possible relationship between CDP and SDPS observations (see discussion).

3. CASE STUDIES ON 24 MARCH 1995 EVENT

On 24 March 1995 the WIND spacecraft (at X $\sim$ 219 R$_E$) observed northward IMF for a prolonged interval

($\sim$18.2 hours, Figure 4 (a)-(c)): During 0110-1924 UT the IMF maintained the northward direction except a short interval during 1233-1239 UT (Figures of solar wind parameters are depicted including the convection delay of $\sim$ 70 min). Especially, the IMF was very northward during both 1-4 UT and 14-15 UT ($B_Z \sim +10$ nT, the latitudinal angle $\theta_B > 70$ deg.). During 4-12 UT and 15-19 UT B_Z decreased slightly but remained strongly northward ($\theta_B > 45$ deg.). While a brief excurtion of B_Z to zero were observed during 1233-1239 UT, any substrorm signatures, such as fast plasma flows, were not observed at the GEOTAIL position.

On the day the GEOTAIL was in the dusk magnetosheath during 0-5 UT, then experienced multiple crossings of the magnetopause during 0354-0716 UT, came into the LLBL at 0716 UT, and exited from the LLBL at 0910 UT [Fairfield et al., 2000]. After 0910 UT the value of B_X oscillated around zero (Figure 4 (d); a black curve), which showed central plasma sheet crossings of the GEOTAIL. Figure 4 (e), V_X, shows stagnant plasma bulk flow in the plasma sheet, which was filled with cold and dense plasma (Figure 4 (f) and (g)). Furthermore, E-t diagram (Figure 4 (i)) shows two-temperature distributions of ions in the plasma sheet. A color bar below in the Figure 4 shows the region observed by the GEOTAIL. The GEOTAIL position in

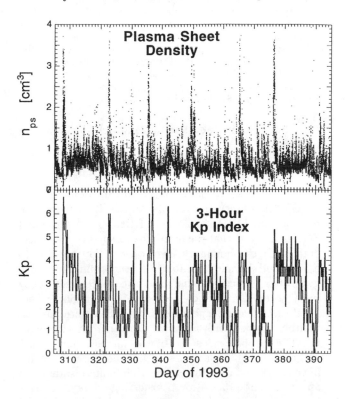

Figure 3. (a) Density in the plasma sheet and (b) Kp plotted for 90 days (November and December 1993 and January 1994) [Borovsky et al., 1997]. The plasma sheet density is measured by three geosynchronous satellites.

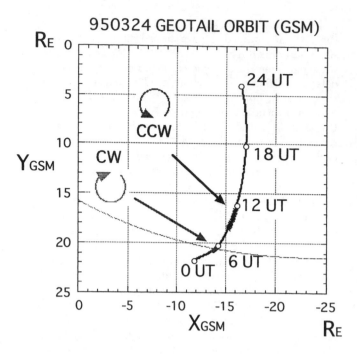

950324 GEOTAIL ORBIT (GSM)

Figure 4. Overview of WIND and GEOTAIL observations on 24 March 1995. From top, (a)-(c) IMF parameters from WIND (Convection delay ~ 70 min are included.), (d) B_X (a black curve) and B_{lobe} (red curves), (e) V_X, (f) ion density from GEOTAIL (a black curve) and solar wind density from WIND (a dotted curve), (g) ion temperature (a black curve) and solar wind kinetic energy (a dotted curve), and E-t diagrams of (h) electrons, (i) ions, and (j) solar wind ions are depicted. (k) Plasma sheet thickness and (l) 'vertical content' (see text) are also shown. A color bar below shows the regions (magnetosheath, magnetopause(MP), LLBL, and CDPS) observed by GEOTAIL.

the GSM coordinate system on 24 March 1995 is shown in Figure 5.

3.1. Implication of Kelvin-Helmholtz Instability

Fairfield et al. [2000] have concluded, comparing the GEOTAIL observation on 24 March 1995 with two-dimentional MHD simulation of this event [Otto and Fairfield, 2000], that the multiple crossings of the spacecraft with the boundary layer on the day are due to a Kelvin-Helmholtz instability at the boundary which generates vortices moving past the spacecraft. They have noted that in the near-equatorial region where northward magnetospheric magnetic fields are approximately aligned with northward magnetosheath fields, GEOTAIL observed some of the largest, most rapid magnetic field fluctuations seen anywhere in space. They have concluded that the Kelvin-Helmholtz instability is an important process for transferring energy, momentum, and particles to the magnetotail during times of very northward IMF.

Recently, Nishino et al. [2002a] have performed a detailed analysis of the rotational sense of the velocity field during the interval same as Fairfield et al. [2000], and found that the "vortical structure" was dominant not only inside but also outside the dusk magnetosphere. Figure 6-(A) shows the observation during 0440-0540 UT including the 40-min interval when the GEOTAIL was located at (−13.9, 20.6, 5.9) R_E. Most of the 40-min interval the satellite was in the magnetosheath, detecting the cold and dense solar wind tailward flow with the velocity of $V_X = -290 \sim -450$ km/s (Figure 6-(A) (a), (e), and (f)). The azimuthal angle of the bulk plasma flow in the average co-moving frame is shown in Figure 6-(A) (d). During 0440-0540 UT the flow vector rotated in the clockwise (CW) direction as viewed from the north with the period of roughly 3-4 min. This sense of rotations was reverse to that observed just inside the magnetosphere (after 0910 UT) on the day (Figure 6-(B), which shows the observation during 0930-1030 UT). The counterclockwise (CCW) sense of rotation within the magnetotail is consistent with the statistical result by Hones et al. [1981]. The position at which these rotations were observed is shown in Figure 5. The red points correspond to clockwise (CW) rotations and the blue counterclockwise (CCW). It is noted that the period of these rotations was roughly 3-4 min both inside and outside.

3.2. Mixed Ion Observation in the Tail-LLBL

In the plasma sheet, in which rotations of bulk flow velocity were observed, mixture of cold and dense plasmas was observed. Fujimoto et al. [1998] have reported that the cold-dense plasma, in this case, is continuously detected as the spacecraft moves inward from the magnetospheric boundary to deep inside the magnetotail. Since the solar wind data showed little dynamic pressure variation during the interval, they have interpreted the long duration of the cold-dense status as indicative of a large spatial extent of the region: The cold-dense plasma is not spatially restricted to a thin layer attached to the magnetopause (tail-LLBL) but contributes an entity occupying a substantial part of the magnetotail.

Three-dimentional electron distributions were obtained after 0648 UT. Fujimoto et al. [1998] have shown that electrons near 145 eV are found to be heated bidirectionally and field-aligned in the magnetosphere-like regions (Figure 7 (a)). After the last fast tailward flow at ~0900 UT, thermal electrons are persistently seen to show clear bidirectional anisotropy up to 300 eV energy range. This feature is seen to fade out after 1300 UT when the ion temperature starts to increase in the most inner part. The observations of the fading out of the bidirectional anisotropy and the heating of the cold-dense ions suggest that the plasma from the flank tends to lose its distinguishing characteristics as they reside in the inner part of the magnetotail being subject to some internal processes.

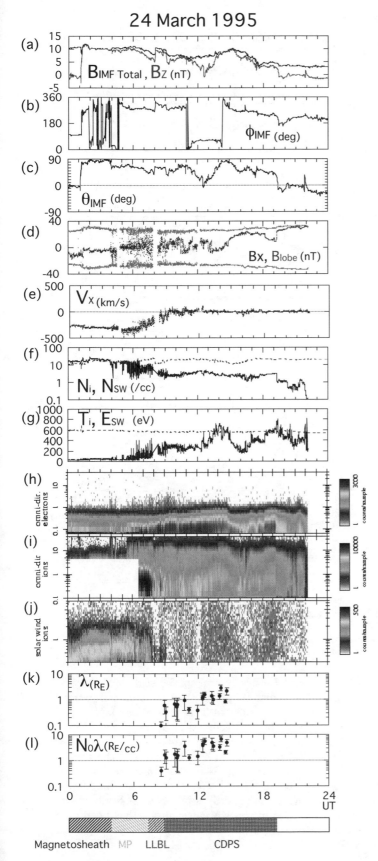

24 March 1995

(a) $B_{IMF\ Total}$, B_Z (nT)

(b) ϕ_{IMF} (deg)

(c) θ_{IMF} (deg)

(d) B_x, B_{lobe} (nT)

(e) $V_{X\ (km/s)}$

(f) N_i, N_{SW} (/cc)

(g) T_i, E_{SW} (eV)

(h) omni-dir. electrons

(i) omni-dir. ions

(j) solar wind ions

(k) $\lambda_{(R_E)}$

(l) $N_0\lambda_{(R_E/cc)}$

Magnetosheath   MP   LLBL   CDPS

The observed ion distribution function was a super-position of cold magnetosheath-like distributions and hotter magnetosphere-like distributions (Figure 7 (b)). The cold-dense ions have only slow tailward convection velocity and are even flowing sunward in some parts. The slow convection and the electron characteristics are taken to be suggestive of the closed topology of the field lines. Charge neutrality seems to be maintained by electrons which are heated bidirectionally. Candidates for transport process of these electrons are (1) an accompanying entry of magnetosheath electrons and (2) a supply from the ionosphere.

3.3. Increase of vertical content

In this 24 March 1995 event Nishino et al. [2002b] also have found that the plasma 'vertical content', namely the product $N_0\lambda$ with the plasma density N_0 at the plasma sheet center and the characteristic thickness λ, showed an order-of-magnitude increase within several hours.

After multiple crossings with the magnetopause the GEOTAIL observed the plasma sheet which was filled with the CDP. Nishino et al. [2002b] have estimated λ and $N_0\lambda$ during 0832 ~ 1445 UT (Figure 4 (k) and (l)) using Sergeev's method which utilizes the oscillation of B_X around zero (see Appendix). Both λ and $N_0\lambda$ gradually increase during the CDP observations. Since variations in N_0 during the interval of 0832-1445 h were within a factor of 2, the order-of-magnitude change of $N_0\lambda$ (0.4 R_E/cc → 6-8 R_E/cc) was due to the increase of λ (0.1 R_E → 2-3 R_E). Nishino et al., [2002b] have shown other two cases of N-IMF periods (9 February 1995 and 4 April 1995 events) in which 'vertical content' increases significantly within several hours. The increase rate of 'vertical content' in these 3 events corresponds to a fairly large plasma transport rate from the solar wind to the near-earth and mid-distant magnetotail during the N-IMF periods, which is estimated to be ~ 10^{26} protons/sec under the assumption that the spatial scale of the near to mid-distant tail current sheet in the X and Y directions are ~50 R_E and ~40 R_E, respectively.

4. DISCUSSION

Cold-dense status of the plasma sheet in the near-earth and mid-distant region likely corresponds to N-IMF condition. While in the recent studies it has been shown that magnetic reconnection occurs during the N-IMF [e.g. Nishida et al., 1998], the effect of this type of reconnection is significant in the distant tail ($X \leq -100$ R_E) and is thought to be difficult to transport solar

Figure 5. GEOTAIL orbit in the XY plane on 24 March 1995. Clockwise (CW) rotations are shown by red points, and counterclockwise (CCW) by bule.

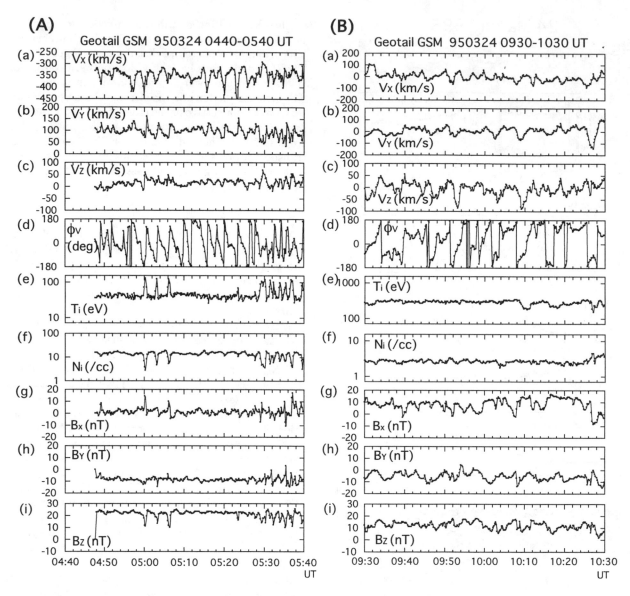

Figure 6.   Clear examples of (A) clockwise (CW) rotations (outside the magnetosphere) and (B) counterclockwise (CCW) rotations (inside). GEOTAIL observations during (A) 0440-0540 UT and (B) 0930-1030 UT of 24 March 1995 are plotted. From top, (a)-(c) 3 components of plasma bulk velocity, (d) azimuthal angle of plasma bulk flow ((A) in the 'co-moving' frame and (B) in the original GSM frame), (e) ion temperature, (f) ion density, (g)-(i) 3 components of magnetic field are shown.

wind plasma into the near-earth plasma sheet. However, recent several studies have suggested that plasma particles which are of magnetosheath origin enter directly into the plasma sheet via near-Earth tail flanks (e.g. Terasawa et al.[1997], Borovsky et al.[1997]). In their cases the mass loading from the solar wind into the plasma sheet appears to be strongest under the N-IMF.

As estimated from the changes of vertical content of the plasma sheet, transport rate during the N-IMF is

$$\sim 10^{26} \left(\frac{L_X}{50\ R_E} \right) \left(\frac{L_Y}{40\ R_E} \right) \text{[/sec]}, \qquad (1)$$

where L_X and L_Y are typical spatial scale of the near-earth and mid-distant plasma sheet. This estimation is consistent with that estimated by Eastman et al. [1985]. Since the transport rate by magnetic reconnection during the southward IMF is estimated to be 10^{27}/sec [Hultqvist et al.,1999], the effect of the transport processes during the N-IMF is not negligible.

Borovsky et al. [1997] have shown that the superdense plasma sheet (SDPS) is observed in conjunction with sharp rises in the Kp index after it has been low for an extended period. We argue that the origin of SDPS in the near-earth plasma sheet can be CDP in

(a)

BCE CUT 950324 08:03:56 - 08:04:58
Electron Phase Space Density (MKS)

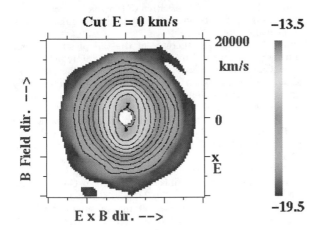

(b)

BCE CUT 950324 08:03:56 - 08:04:58
Ion Phase Space Density (MKS)

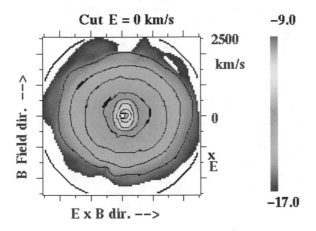

Figure 7. An example of particle distribution functions in the tailward flowing flux tubes. (a) Football shaped bidirectional electron distribution. (b) Ions showing two-component feature. (From Fujimoto et al., [1998], Plate 9 (b) and (a). The order of panels are changed from that in the original paper.)

the mid-distant plasma sheet. A possible scenario of transport process of the CDP in the plasma sheet is as follows: If the IMF turns northward for several hours during the slow and dense solar wind phase, the plasma sheet in the region of X = −15 to −50R$_E$ becomes cold and dense. When the IMF turns southward, substorm activity commences and the enhanced earthward convection carries the dense plasma inward to be observed as SDPS in the inner magnetosphere.

As reviewed in this article, recent observations of the plasma sheet strongly imply a direct entry of plasma

particles from the solar wind into the near-earth and mid-distant plasma sheet via the LLBL under the N-IMF conditions. Furthermore, the fact that under the N-IMF conditions 'vertical content' increases gives rise to a new question; What is the balance between supply and loss process of plasma particles in the plasma sheet during the N-IMF intervals? The supply process of the CDP into the plasma sheet as well as the loss process of the CDP during the N-IMF should be studied in future works.

APPENDIX A: SERGEEV'S METHOD

As a useful tool to measure the current sheet thickness, λ, Sergeev et al. [1998] have proposed a new method based on the correlation analysis between the normalized time variation of the X component of the magnetic field, $\dot{B}_X/B_{X\text{lobe}}$ and Z component of the plasma bulk velocity, V_Z: When there is significant negative correlation between $\dot{B}_X/B_{X\text{lobe}}$ and V_Z, λ is given as the coefficient of the regression line, $V_Z \sim -\lambda \dot{B}_X/B_{X\text{lobe}}+$(offset errors).

Nishino et al. [2002b] have applied this Sergeev's method to the plasma sheet data from the GEOTAIL measurement. In Nishino et al. [2002b], instantaneous values of the magnetic field $|B_{X\text{lobe}}|$ in the lobe (Figure 4 (d); red curves) are calculated first under the assumption of the one-dimensional pressure balance. Next, high-β intervals of 2-5 min duration were chosen by posing the condition that $|B_X/B_{X\text{lobe}}|$ was continuously smaller than 0.3. For each of these intervals Nishino et al. [2002b] made a regression analysis between $\dot{B}_X/B_{X\text{lobe}}$ and V_Z, and chose those whose correlation coefficient was negative and significant (≤ -0.6). Among 2-5 min intervals relating to the same plasma sheet event one with the best negative correlation was chosen. It was found that the average of λ in the region $-15 \leq X \leq -50$ R$_E$ when the magnetosphere is quiet (Kp $\leq$ 2+) is 2 $\sim$ 3 R$_E$ [Nishino, 2000], which is close to that obtained by Pulkkinen et al. [1993] ($\sim$ 3R$_E$) in the distant tail.

Acknowledgments. This research is partially supported by ACT-JST (Research and Development for Applying Advanced Computational Science and Technology of Japan Science and Technology Corporation).

REFERENCES

Baker, D. N., et al., The role of heavy ionospheric ions in the localization of substorm disturbances on March 22, 1979 - CDAW 6, *J. Geophys. Res., 90,* 1273-1281, 1985.

Borovsky, J. E., et al., The superdense plasma sheet: Plasmasheric origin, solar wind origin, or ionospheric origin?, *J. Geophys. Res., 102,* 22,089-22,097, 1997.

Chen, et al., Anomalous aspects of magnetosheath flow and of the shape and oscillations of the magnetopause during an interval of strongly northward interplanetary magnetic field, *J. Geophys. Res., 98,* 5727-5742, 1993.

the inner edge of the low latitude boundary layer produced by these B_y reconnection flows across the dayside.

Acknowledgment. This work has been supported over the past decade by several awards from the National Science Foundation. At the present time this work is supported by NSF Grants ATM - 9628706 and OPP - 9876473. Numerous colleagues and students have contributed to this research over the years and they are represented in the bibliographic citations.

REFERENCES

Clauer, C., M. McHenry, and E. Friis-Christensen, Observations of filamentary field-aligned current coupling between the magnetospheric boundary layer and the ionosphere, in *Physics of Magnetic Flux Ropes*, edited by C. Russell, p. 565, AGU, Washington, D.C., 1990.

Clauer, C. R., Solar wind - magnetosphere - ionosphere coupling: Dayside observations of low latitude boundary layer waves, in *Physics of Space Plasmas (1998)*, edited by a. J. R. J. T. Chang, MIT Center for Theoretical Geo/Cosmo Plasma Physics, Cambridge, MA, 1998.

Clauer, C. R., and A. J. Ridley, Ionospheric observations of magnetospheric low latitude boundary layer waves on August 4, 1991, *J. Geophys. Res.*, *100*, 21873–21884, 1995.

Clauer, C. R., A. J. Ridley, R. J. Sitar, H. J. Singer, A. S. Rodger, E. Friis-Christensen, and V. O. Papitashvili, Field-line resonant pulsations associated with a strong dayside ionospheric shear convection flow reversal, *J. Geophys. Res.*, *102*, 4585–4596, 1997.

Fejer, J., Hydromagnetic stability at a fluid velocity between compressible fluids, *Phys. Fluids*, *7*, 499, 1964.

Friis-Christensen, E., M. A. McHenry, C. R. Clauer, and S. Vennerstrøm, Ionospheric travelling convection vorticies observed near the polar cleft: A triggered responce to sudden changes in the solar wind, *Geophys. Res. Lett.*, *15*, 253–256, 1988a.

Friis-Christensen, E., Vennerstrøm, C. Clauer, and M. A. McHenry, Irregular magnetic pulsations in the polar cleft caused by travelling ionospheric current vortices, *Adv. Space Res.*, *8*, 311, 1988b.

Glassmeier, K., M. Hönisch, and J. Untiedt, Ground-based and satellite observations of travelling magnetospheric convection twin vorticies, *J. Geophys. Res.*, *94*, 2,520, 1989.

Glassmeier, K.-H., Traveling magnetospheric convection twin-vortices: Observations and theory, *Ann. Geophysicae*, *10*(5), 547, 1992.

Glassmeier, K.-H., and C. Heppner, Traveling magnetospheric convection twin vortices: Another case study, global characteristics, and a model, *J. Geophys. Res.*, *97*, 3,977, 1992.

Greenstadt, E. W., J. V. Olson, P. D. Loewen, H. J. Singer, and C. T. Russell, Correlation of pc3, 4 and 5 activity with solar wind speed, *J. Geophys. Res.*, *84*, 6,694, 1979.

Heelis, R., W. Hanson, and J. Burch, Ion convection velocity reversals in the dayside cleft, *J. Geophys. Res.*, *81*, 3,803, 1976.

Heikkila, W., T. Stockflet-Jorgensen, L. J.Lanzerotti, and C. G. Maclennan, A transient auroral event on the dayside, *J. Geophys. Res.*, *94*, 15291, 1989.

Junginger, H., and W. Baumjohann, Dayside long period magnetospheric pulsations: Solar wind dependence, *J. Geophys. Res.*, *93*, 877, 1988.

Kokubun, S., N. Erickson, T. A. Fritz, and R. L. McPherron, Local time asymmetry of pc4-5 pulsations and associated particle modulations at synchronous orbit, *J. Geophys. Res.*, *94*, 6607, 1989.

Lanzerotti, L., A. Wolfe, N. Trivedi, C. Maclennan, and L. Medford, Magnetic impulse events at high latitudes: Magnetopause and boundary layer plasma processes, *J. Geophys. Res.*, *95*, 97, 1990.

Lühr, H., and W. Blawert, Ground signature of travelling convection vortices, in *Solar Wind Sources of Magneospheric Ultra-Low-Frequency Waves*, edited by M. Engebretson, K. Takahashi, and M. Scholer, p. 231, AGU, Washington, D.C., 1994.

McHenry, M. A., and C. R. Clauer, Modeled ground magnetic signatures of flux transfer events, *J. Geophys. Res.*, *92*, 11,231, 1987.

McHenry, M. A., C. R. Clauer, E. Friis-Christensen, and J. D. Kelly, Observations of ionospheric convection vortices: Signatures of momentum transfer, *Adv. Space Res.*, *8*, 315, 1988.

McHenry, M. A., C. R. Clauer, and E. Friis-Christensen, Relationship of solar wind parameters to continuous dayside, high latitude traveling ionospheric vortices, *J. Geophys. Res.*, *95*, 15,007, 1990a.

McHenry, M. A., C. R. Clauer, E. Friis-Christensen, P. T. Newell, and J. D. Kelly, Ground observations of magnetospheric boundary layer phenomena, *J. Geophys. Res.*, *95*, 14,995, 1990b.

Miura, A., Kelvin-Helmholtz instability at the magnetospheric boundary: Dependence on the magnetosheath sonic Mach number, *J. Geophys. Res.*, *97*, 10665, 1992.

Newell, P., and C.-I. Meng, Mapping the dayside ionosphere to the magnetosphere according to particle precipitation characteristics, *Geophys. Res. Lett.*, *19*, 609, 1992.

Newell, P., W. Burke, E. Sanchez, C.-I. Meng, M. Greenspan, and C. Clauer, The low-latitude boundary layer and the boundary plasma sheet at low altitude: Prenoon precipitation regions and convection reversal boundaries, *J. Geophys. Res.*, *96*, 21,013, 1991.

Ridley, A. J., and C. R. Clauer, Characterization of the dynamic variations of the dayside high-latitude ionospheric convection reversal boundary and relationship to interplanetary magnetic field orientation, *J. Geophys. Res.*, *101*, 10,919, 1996.

Sckopke, N., G. Paschmann, G. Haerendel, B. U. O. Sonnerup, S. J. Bame, T. G. Forbes, J. E. W. Hones, and C. T. Russell, Structure of the low-latitude boundary layer, *J. Geophys. Res.*, *86*, 2099–2110, 1981.

Sitar, R. J., J. B. Baker, C. R. Clauer, A. J. Ridley, J. A. Cumnock, V. O. Papitashvili, J. Spann, M. J. Brittnacher, and G. K. Parks, Multi-instrument analysis of the ionospheric signatures of a hot flow anomaly occurring on July 24, 1996, *J. Geophys. Res.*, *103*, 23,357, 1998.

Southwood, D., The hydrodynamic stability of the magnetospheric boundary, *Planet. Space Sci.*, *16*, 587–605, 1968.

Southwood, D., Magnetopause Kelvin-Helmholtz instability, in *Proceedings of Magnetospheric Boundary Layers Conference, Eur. Space Agency Spec. Publ. ESA SP 148*, p. 357, European Space Agency, 1979.

C. R. Clauer, Space Physics Research Laboratory, University of Michigan, 2455 Hayward, Ann Arbor, MI 48109-2143. (e-mail: bob.clauer@umich.edu)

Anti-Parallel Reconnection at the Dayside Magnetopause: Ionospheric Signatures and Implications for the low Latitude Boundary Layer

A. S. Rodger, G. Chisham, I. J. Coleman, M. P. Freeman and M. Pinnock

British Antarctic Survey, Madingley Road, Cambridge, United Kingdom.

Reconnection at the dayside magnetopause is of fundamental importance in determining the dynamics and composition of the magnetosphere and the high latitude ionosphere. There are two competing hypotheses for such reconnection, sub-solar and anti-parallel. In this paper, evidence is provided that suggests that anti-parallel reconnection occurs, at least under some circumstances. Some of the consequences of anti-parallel reconnection are considered. These include the absence of a low latitude boundary layer during southward IMF, the relative timing of reconnection events as observed in the ionosphere, the time-dependence of ion outflow and the implications for the composition of the magnetosheath, and the transient nature of reconnection. Some suggestions for tests that would allow differentiation between the anti-parallel and sub-solar reconnection are also made.

1. INTRODUCTION

Reconnection on the dayside magnetopause is the major process by which energy, momentum and mass are transferred from the shocked solar wind into the magnetosphere. Reconnection also provides an important mechanism by which magnetosheath and magnetosphere plasmas mix, and hence contributes significantly to the formation of the low latitude boundary layer. There is still considerable debate as to whether reconnection at the dayside magnetopause occurs near the sub-solar region [*Gonzalez and Mozer,* 1974] or where the magnetosheath and magnetospheric fields are anti-parallel [*Crooker,* 1979]. The large-scale survey of flux transfer events on the magnetopause by *Kawano and Russell* [1997] tends to support the sub-solar hypothesis whereas two spacecraft observations by *Safrankova et al.* [1998] and the statistical survey of energetic particle precipitation in the cusp by *Newell et al.* [1995] support the anti-parallel hypothesis.

Earth's Low-Latitude Boundary Layer
Geophysical Monograph 133

These two models will give several different spatial and temporal characteristics of the low latitude boundary layer, the magnetosheath, and the ionospheric signatures of these processes. It is essential to understand these features and their time dependence to explore further the physics of the reconnection process itself – one of the major unresolved topics in solar, solar terrestrial and astro physics.

The conflicting evidence indicates one of two basic problems. First the observations to date do not allow definitive discrimination between the two processes. Alternatively, it is possible that both sub-solar and anti-parallel reconnection occur, but that the time history of solar wind-magnetosphere interactions dictates which reconnection scenario is more likely. Modelling developed by *Coleman et al.* [2001] provides a critical test that allows differentiation of the two reconnection hypotheses. Under extreme dipole tilt conditions, these authors predict a unique ionospheric convection signature of anti-parallel reconnection, that was subsequently identified in data from HF SuperDARN radars [*Greenwald et al.,* 1995].

The purpose of this paper is to provide an overview of the key findings of the *Coleman et al.* [2001] work, and then discuss briefly some of the consequences for the low latitude boundary layer and its ionospheric footprint. The effects discussed involve ion outflow, the timing of the

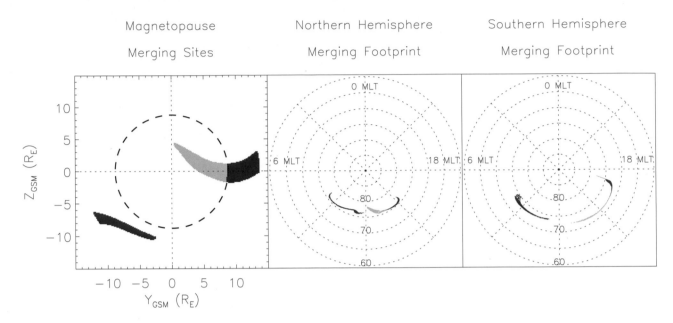

Figure 1. The left hand panel shows the location on the magnetopause in the y-z plane where sub-solar reconnection is expected for December solstice for IMF Bz =-3 nT, and By = 3nT. The dashed circle shows the Alfvénic boundary. This is defined using the approach of *Cowley and Owen* [1989] but see *Coleman et al.* [2000] for more details. The right hand panels show where the reconnection footprints map to the ionosphere using the *Tsyganenko* [1996] model in the northern (middle) and southern (left) hemispheres (after Coleman et al. 2001). The dark and lightly shaded regions correspond to the super-Alfvénic and sub-Alfvénic regions respectively on the left hand panels.

onset of flows and electric fields in the ionosphere and the low latitude boundary layer, and implications for transient reconnection. We shall illustrate some of the consequences using published examples.

2. MAPPING RECONNECTION SITES FROM THE MAGNETOPAUSE TO THE IONOSPHERE

2.1. Sub-Solar Reconnection

The Tsyganenko magnetospheric field model [*Tsyganenko*, 1995, 1996; *Tsyganenko and Stern*, 1996] is a good model for mapping the magnetopause to the ionosphere, especially on the dayside where the field line stretching is not important. The x-line can be readily mapped to the ionosphere once the dynamic pressure, year, day and UT are defined. An example for December solstice at 1700 UT is shown in Figure 1, i.e. when the dipole tilt is at a minimum (-34°). In the sub-solar reconnection case, the key feature at the magnetopause is that reconnection occurs along the z=0 axis. When the region is mapped to the ionosphere, some differences can be seen between the hemispheres. The sub-Alfvénic region (light shading) on the magnetopause maps to about 7 h of magnetic local time (MLT) in the winter (north) but less than half this in the south (summer). As the same potential

difference is applied to both reconnection x-lines in the ionosphere, the flow velocity should be substantially lower in the winter hemisphere. Some support for this suggestion has been provided by *Pinnock et al.* [1999]. Also the latitude of the ionospheric footprint of reconnection is about 2-3° higher in the summer hemisphere near noon, consistent with the observations of *Newell and Meng* [1989].

2.2. Anti-Parallel Reconnection

Coleman et al. [2000] developed a model to identify regions on the magnetopause where the magnetosheath and the magnetosphere magnetic fields were anti-parallel. They used the Tsyganenko magnetic field model for the magnetosphere field. For the magnetosheath field, they assumed perfect draping of the interplanetary magnetic field - a reasonable approximation on the dayside but one that becomes progressively less good tailward of the dawn-dusk flanks. 'Anti-parallel' was defined as being where the magnetic fields were oppositely directed to within 10°. This assumption is the same as that of *Luhmann et al.* [1984]; in practice the results presented here are insensitive to the precise definition of 'anti-parallel'.

Figure 2 shows the corresponding example for anti-parallel reconnection for the same conditions and in the

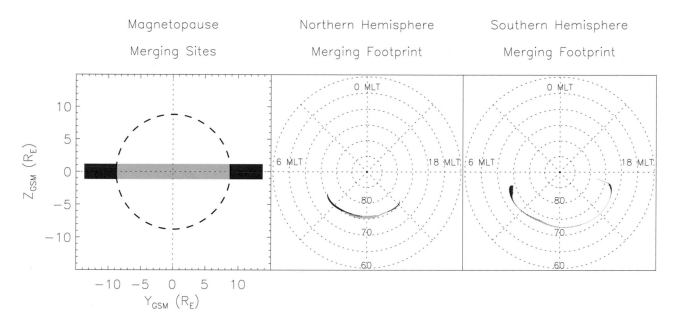

| Magnetopause Merging Sites | Northern Hemisphere Merging Footprint | Southern Hemisphere Merging Footprint |

Figure 2. The same as figure 1 but for the anti-parallel case [after *Coleman et al.*, 2001].

same format as Figure 1. There are some very significant contrasts. On the magnetopause, there are two reconnection sites separated by >10 Earth radii (Re). In the morning sector, a significant proportion of the anti-parallel reconnection region falls within the sub-Alfvénic regime and lies near the equatorial plane. However the afternoon sector region lies completely in the super-Alfvénic sector and well above the equatorial plane. When these regions are mapped to the summer (southern) ionosphere, the extent and latitude are rather similar to the sub-solar pattern (see Figure 1) though there are small but noticeable differences near noon. However there are very large differences from the sub-solar case in the winter (northern hemisphere). The morning and afternoon x-lines are separated about noon by more than 1 hour. The separation of the two x-lines in the ionosphere maximises when the dipole tilt is at its extreme value, and when $|By| \approx Bz$ and Bz is negative. The dipole tilt causes the reconnection sites to move progressively towards more positive Z positions on the magnetopause from equinox to northern winter solstice. Conversely the reconnection sites are further below the magnetic equator as the southern winter solstice is approached. Through the course of a day, the reconnection sites migrate in the z direction, but to a lesser extent than the seasonal variation again due to the offset of the magnetic and geographic poles.

If a potential difference is applied to the reconnection x-lines in the ionosphere, then a fundamental consequence is that there must be flow equatorward between the two reconnection sites in the winter hemisphere for the anti-parallel case, provided that the polar cap boundary is expanding i.e. as occurs in a growth phase. The flow in the corresponding location near noon for the sub-solar case is poleward. This difference between the two cases allows a definitive test that can separate the sub-solar from the anti-parallel reconnection hypotheses. Further justification for the assumptions made here is provided in *Coleman et al.* [2001] and the references therein.

To test the hypothesis against observations, it is necessary to find periods when the IMF is southward, By is large and both components are steady for several hours near solstice, as well as there being large-scale, 2-D maps of ionospheric convection available from near noon. Only a few examples that satisfy these stringent conditions have been found and all of them show the same general characteristics i.e. a significant component of equatorward flow near noon with strong poleward flow component either side of noon. *Coleman et al.* [2001] published an example for 10 December 1997. The solar wind and the IMF both remained steady for ~5 hours. The SuperDARN HF radars [*Greenwald et al.*, 1995] provided excellent 2-D imaging of ionospheric, convection at the appropriate latitude. This key feature, illustrated in Figure 3, is a region about 1 hour wide centred on magnetic noon where the flow has a significant equatorward component (~300 ms^{-1}), even though Bz was steady at –10 nT and By=10 nT. This direction is not that expected from the sub-solar hypothesis and in contrast to the conventional wisdom of flow into the polar cap near noon as outlined by *Cowley and Lockwood* [1992] amongst others but the observation is consistent with the anti-parallel reconnection hypothesis described above.

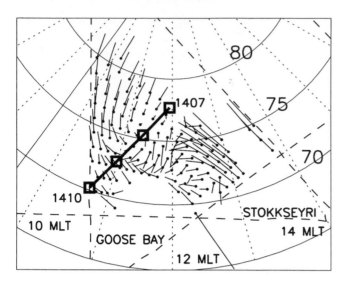

Figure 3. 2-D velocity vectors determined from the Goose Bay and Stokkseyri SuperDARN HF radars. Line-of-sight velocity data from both radars were averaged over a 20-minute interval (1340-1400 UT) on 10 December 1997. The track of a concurrent DMSP F14 satellite pass is shown. The equatorward edge of the cusp precipitation is observed at 71.25° but there is a second ion dispersion event at 73.0°, consistent with arising from a second x-line.

Although the evidence from the example described above, and a few other case studies [*Maynard et al.*, 2001] provide substantive evidence for anti-parallel reconnection to be dominant process occurring, it is by no means proof that anti-parallel reconnection is invariably dominant. Indeed in an as-yet unpublished study, Chisham et al. propose that the time history of reconnection may be an important consideration over the relative importance of sub-solar and anti-parallel reconnection.

Sub-solar and anti-parallel reconnection may occur simultaneously. Optical observations from Svalbard show evidence of sheath-like electron precipitation poleward and equatorward of the station at the same time [*Sandholt et al.*, 1998]. All their examples occur during large dipole tilt conditions, as only around the December solstice is Svalbard in darkness near noon. However the examples could be simply a result of time-dependence. For example, during a transition from northward to southward IMF, reconnection will start at a low latitude site ~5-10 minutes before the new orientation of the IMF reaches the high-latitude site causing reconnection to cease there. Therefore it is important to examine both the time dependence of the bifurcated optical signatures and the direction of motion of any transients to determine whether the observed signatures are a new configuration of the reconnection sites emerging, or a manifestation of the sub-solar and anti-parallel reconnection occurring simultaneously.

3. SOME CONSEQUENCES OF RECONNECTION FOR THE BOUNDARY LAYER AND THE IONOSPHERE

In this section, we explore some consequences of reconnection both for the formation of the boundary layer and the ionospheric signatures. Specifically we consider whether some of the existing observations can be explained by anti-parallel or sub-solar reconnection hypothesis.

3.1. An Absence of a LLBL during Southward IMF

Eastman et al. [1996] reported that on about 10% of occasions that the ISEE 2 and AMPTE spacecraft crossed the dayside magnetopause near noon, there was no evidence of a low latitude boundary layer (LLBL). Over half of these examples occurred when the IMF was southward. No substantive explanation was offered then, but the anti-parallel hypothesis can provide an explanation. The shaded regions in Figure 4 show schematically where reconnection can occur on the dayside magnetopause when IMF Bz is negative and By positive in the anti-parallel

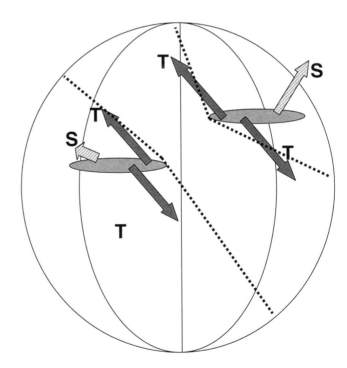

Figure 4. The shaded region show schematically where anti-parallel reconnection is likely to occur on the dayside magnetopause for IMF Bz southward and By positive with a significant dipole tilt (compare with figure 2). The arrows mark the direction that the magnetic tension (T) and sheath (S) forces act and the subsequent motion of a reconnected flux tube will be the vector addition of T and S. The region enclosed by the broken lines is where there will be no reconnected flux overlying the dayside magnetopause.

case. The subsequent motion of the reconnected flux tubes is controlled by the effects of sheath flow (S) and tension (T), the latter depending on the orientation of IMF By. This is shown in Figure 4 by arrows. The net motion of the reconnected flux tube is the vector addition of the sheath and tension forces, the latter being different for the flux tubes connected to the northern and southern hemispheres. The broken line on the figure marks regions where reconnected flux tubes will not overly the magnetopause for the prevailing IMF conditions. There is a substantive sub-solar region stretching from near noon to the dusk flank where there would be no reconnected flux and hence no evidence of an open low latitude boundary layer. The shape and size of this region will depend upon the orientation of the IMF.

If one uses a similar approach but with the sub-solar model (Figure 1) there will be no regions near noon where recently opened flux tubes will be absent. Hence, the unusual magnetopause crossing identified by *Eastman et al.* [1996] could be consistent with the anti-parallel reconnection hypothesis.

3.2. Timing of 'Events'

The distance between the afternoon reconnection location in Figure 2 and the northern ionosphere is about 10 Re shorter in distance compared with the corresponding distance to the southern ionosphere. Hence the arrival of the Alfvén wave in the ionosphere that carries the information about reconnection in the north will be several minutes earlier than in the south. *Chisham et al.* [2000] provide an example of this effect by observing a hemispheric difference in the response time of ionospheric flows to a change in the IMF; they show a reconnection event was most likely to occur about 6 Re above the equatorial plane. If reconnection occurs at the sub-solar point, then the time delay to the two hemispheres will be approximately the same. Hence this is further evidence reconnection occurs where the fields are approximately anti-parallel. *Chisham et al.* [2000] suggest that the IMF Bx component plays an important role in determining the location of the reconnection site, but Figure 2 only considers effects in the y- and z planes. In reality, all three dimensions are important.

By applying the same principles about distance between the reconnections sites and the ionosphere, and using Figure 2 as an example, the afternoon reconnection events will be seen several minutes before those in the morning sector in the northern ionosphere. This assumes that reconnection occurred simultaneously at the two locations, an assumption that would seem unlikely as the phase front of the IMF would need to be normal to the x direction – a rare occurrence. Observations consistent with this scenario

have been presented by *Maynard et al.* [2001] for a tilted IMF phase front. In general, the timing of events in the morning and afternoon ionospheres will be a convolution of the temporal characteristics of the IMF and the location of reconnection events on the magnetopause. As the three dimensional structure of the magnetosheath fields become better understood through the exploitation of Cluster data, the relative timing of reconnection signatures in the morning and afternoon sectors will be a more important diagnostic for the location of reconnection sites on the magnetopause. The points are made assuming that no draping occurs, but the range of geophysical conditions under which this assumption is valid has yet to be determined.

Energetic ion dispersion is a well-known characteristic of recently-reconnected flux tubes [*Smith and Lockwood*, 1996]. It results from the progressively longer time of flight of the lower energy magnetosheath ions reaching the ionosphere. During their transit from the magnetopause the ions convect, the direction and magnitude depending mainly upon combination of forces imposed by field line tension and sheath flow. As illustrated in Figure 4, the motion of the morning and afternoon sector reconnected flux tubes in the two hemispheres will not be identical as the tension and sheath forces are substantially different. Onc consequence is that the location where the most energetic ions crossing the magnetopause as a result of reconnection reach the ionosphere will be different for the morning and afternoon sectors owing to the different length of flux tubes from the reconnection sites (see Figure 5). The displacement (d) of the open/closed field line boundary from the equatorward limit of its ionospheric proxies is discussed in more detail by *Rodger* [2000]. The orientation and time evolution of ion dispersion events arising from the morning and afternoon reconnection lines will also be different, also illustrated in Figure 5. A polar-orbiting satellite will detect a discontinuity in the ion energy spectrum as it crosses from ion dispersion event exiting from the morning x-line into that of the afternoon x-line. *Trattner et al.* [2001] have provided substantial evidence that such structures not only exist but also persist for many hours provided that the IMF remains stable. From the energetic particle data alone, it would not be possible to determine whether the spacecraft has passed through a spatial or temporal feature. In the recent literature, discontinuities in cusp ion spectra have tended to be considered only temporal variations of reconnection [*Smith and Lockwood*, 1996] but this new interpretation suggests that the events could be spatial. Snapshots of the ion dispersion events taken 1-2 minutes taken by two DMSP spacecraft on very similar trajectories show provide some evidence for time-dependent multiple x-line reconnection [*Boudouridis* et al., 2001]. Further examples are likely to provide important evidence about reconnection.

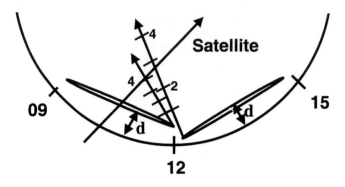

Figure 5. The circular line represents the ionospheric footprint of the open-closed field line boundary in the northern hemisphere. The extended ellipses are where the most energetic sheath ions that crossed the magnetopause at the onset of reconnection impact the ionosphere. The trajectories of two flux tubes that have recently reconnected, one each from the morning and afternoon x-lines with the tick marks indicating the time in minutes since the most energetic ions resulting from reconnection impacted the ionosphere. A typical trajectory of a polar-orbiting satellite, such as DMSP, is shown. The satellite will detect a discontinuity in the ion spectra it moves from the morning to the afternoon ion dispersion spectra.

3.3. Time Dependence of ion Outflow

A critical diagnostic of the open low latitude boundary layer is the observation of significant fluxes of singly-charged oxygen outside the magnetopause current layer [*Onsager et al.*, 2001; *Fuselier et al.*, 2001]. However it will take some time for O^+ to reach the observing spacecraft depending upon the field-aligned speed of the O^+ and the distance between the reconnection site and the measurement location. As discussed previously, the separation of the possible reconnection sites can be ±10 Re depending whether a sub-solar or anti-parallel reconnection model is used. This value is determined by comparing the reconnection regions in Figures 1 & 2. The velocity of O^+ along the flux tube is typically a few km s^{-1} at 1-2 Re above the Earth's surface [*Schunk*, 2000] but further acceleration along the flux tube can occur so that O^+ can reach velocities of 100 km s^{-1}. The time taken for O^+ to travel 10 Re is about 10 minutes. Thus the presence of O^+ in the magnetosheath is only a reliable indicator of reconnection sometime after it has occurred. The corollary is that the absence of O^+ is not a definitive diagnostic of a field line that has both ends on the Sun.

The mechanisms that drive the ion outflow in the cusp are still not agreed. *Schunk* [2000] provides a theoretical summary of the principal processes that may be involved. *Øieroset et al.* [2000] have shown that ion conics dominate ion outflow near noon, at least at Viking altitudes (~1-2 Re above the Earth's surface). These fluxes are greater for southward IMF conditions and also when AE is larger.

Somewhat surprisingly the outflow appears to peak near 78° magnetic latitude, which is several degrees higher latitude than the ionospheric footprint of the cusp under the same conditions. Thus where and when the ion outflow crosses the magnetopause may require further study. Cluster data should be able to provide the necessary evidence.

3.4. Transient Reconnection

After more than a decade of study, there is still considerable debate over the relative importance of the temporal and spatial variations of reconnection [see *Pinnock et al.*, 1995; *Lockwood and Davis*, 1996 and the references therein]. In practice, both spatial and temporal reconnection rate variations are likely to occur but the important issue, to which there is no answer yet, is the geophysical conditions that control the reconnection rate. *Rodger et al.* [2000] used the evidence from *Coleman et al.* [2000] to suggest that reconnection must be transient where the sheath flow is super-Alfvénic. The fundamental reason for this conclusion is that to be able to conserve mass across the reconnection x-line in the rest frame of the event, the x-line itself must be propagating anti-sunward, and therefore the necessary conditions required to maintain reconnection cannot be sustained indefinitely along the flanks and into the tail of the magnetosphere. The hypothesis included in *Rodger et al.* [2000] does not address the stability of reconnection within the sub-Alfvénic region; this is a topic for much further research.

In the original hypothesis [*Rodger et al.*, 2000], simple hydro-dynamic flow of the shocked solar wind around the magnetopause was assumed, but now two important caveats must be added. For northward IMF, a depletion layer may form around the magnetopause surface and the Alfvén speed is much higher than would have otherwise been expected [*Fuselier et al.*, 2000]. Therefore it may be possible for the anti-parallel reconnection condition for northward IMF to be satisfied within the sub-Alfvénic region, a factor that was not included in the original hypothesis. A critical test will be to determine the stability of the x-line under northward IMF conditions, possibly using optical data [*Sandholt et al.*, 1998], when simultaneous sheath measurements are available, such as those from Cluster.

The second development of the original hypothesis concerns the fact that the magnetic field can be weak towards the flanks of the magnetosphere near the equatorial plane. This implies a reduction in the local Alfvén speed and hence will result in a distortion of the circles shown in Figures 1 & 2. This region has been called the sash and its location and orientation is dependent upon the IMF [*Siscoe et al.*, 2001].

Figure 2 demonstrates for southward IMF that the reconnection site in the afternoon must be transient

the 55
ager. 1
behavi
which
the ion
May
monize
ionosph
mined
The dat
merging
field. T
ceeds n
latitudes
every m
auroral
holt et a
plasma
response
variation
of such
ine varia
emission
When
tudes, an
to move
al., 2000
that is tie
drapes o
dusk by
reversed
Plate
rocket an
dominate
B_Z) [*Ma*
show the
magnetic
give the r
were add
of the cu
the conne
separator
in the Sou
in the top
IEF phase
during the
the phase
ern Hemis
yellow da
recently o
sult of mei

m
int
sou
bif
(IE
spl
dif
dic
He
an
top
and
Sou
Noi
is d
witl
plai
for
con
of i
ope
ary
thic
the
sma

Earth's Low-Latitude Bou
Geophysical Monograph
Copyright 2003 by the Ai
10.1029/133GM32

whereas most of the morning reconnection line could be stable. Therefore the occurrence of transient reconnection should have a significant local time asymmetry, for southward IMF in the present of a moderate or strong By field at large dipole tilt angles, if the anti-parallel condition is satisfied frequently. This would imply that the temporal nature of the LLBL would be more variable in the afternoon sector for By positive conditions. This asymmetry would not occur for sub-solar reconnection. Some studies of the occurrence of the ionospheric signatures of flux transfer events have been carried out. *Provan et al.* [1999] found some By dependence to the occurrence of the ionospheric signatures of flux transfer events but in a larger study *McWilliams et al.* [2000] did not. Neither study examined the occurrence of FTEs in such a way as to provide a critical test of the sub-solar or anti-parallel reconnection hypotheses.

4. SUMMARY

Coleman et al. [2001] have provided critical evidence that anti-parallel rather than sub-solar reconnection can occur, at least during conditions of strong IMF By and large dipole tilt angles.

There are several important consequences of this finding that affect the formation and characteristics of the low latitude boundary layer. It provides a plausible explanation for:

why *Eastman et al.* [1996] found no LLBL on 10% of satellite passes through the noon magnetopause during southward IMF conditions;

why *Trattner et al.* [2001] found stable but structured ion dispersion events lasting many hours in the cusp;

why there should be an asymmetry of ionospheric transients occurrence about noon depending upon the orientation of IMF By and season;

why an absence of singly-charged oxygen in the magnetosheath does not imply a closed boundary layer owing to the time taken for O^+ to propagate along magnetic flux tubes.

Although there is significant evidence for anti-parallel reconnection, these findings do not necessarily imply that it is the dominant process. Sub-solar reconnection is not excluded. The next step is to determine under what geophysical conditions sub-solar or anti-parallel reconnection is the dominant process by exploiting new data sets such as Cluster, and some of the ideas presented here.

REFERENCES

Baker, K. B., J. R. Dudeney, R. A. Greenwald, M. Pinnock, P. T. Newell, A. S. Rodger, N. Mattin, and C. -I. Meng, HF-radar signatures of the cusp and low latitude boundary layer, *J. Geophys. Res.*, 100, 7671-7695, 1995.

Boudouridis, A., H. E. Spence and T. G. Onsager, Investigation of magnetopause reconnection models using two co-location low altitude satellites: 1. Unifying reconnection geometry, *J. Geophys. Res.*, (submitted).

Chisham, G., M. Pinnock, A. S. Rodger and J. -P. Villain, High-time resolution conjugate SuperDARN radar observations of the dayside convection response to changes in IMF By, *Ann. Geophys.*, 18, 191-201, 2000.

Coleman, I. J., M. Pinnock and A. S. Rodger, The ionospheric footprint of anti-parallel merging regions on the dayside magnetopause, *Ann. Geophys.*, 18, 511-516, 2000.

Coleman, I. J., G. Chisham, M. Pinnock and M. P. Freeman, An ionospheric convection signature of anti-parallel reconnection, *J Geophys. Res.*, (in press).

Cowley, S. W. H. and C. J. Owen, A simple illustrative model of open flux tube motion over the dayside magnetopause, *Planet. Space Sci.*, 37, 1461-1475, 1989.

Cowley, S. W. H., and M. Lockwood, Excitation and decay of solar wind-driven flows in the magnetosphere-ionosphere system, *Ann. Geophys.*, 10, 103-115, 1992.

Crooker, N. U., Dayside merging and cusp geometry, *J. Geophys. Res.*, 84, 951- 959, 1979.

Eastman, T. E., S. A. Fuselier and J. T. Gosling, Magnetopause crossings without a boundary layer, *J. Geophys. Res.*, 101, 49-57, 1996.

Fuselier, S. A., S. M. Petrinec and K. J. Trattner, Stability of the high-Latitude reconnection site for steady northward IMF, *Geophys. Res. Lett.*, 27, 473-476, 2000.

Fuselier et al. 2001 – this volume.

Gonzalez, W. D. and F. S. Mozer, A quantitative model for the potential resulting from reconnection with an arbitrary interplanetary magnetic field, *J. Geophys. Res.*, 79, 4186- , 1974.

Greenwald, R. A., K. B. Baker, J. R. Dudeney, M. Pinnock, T. B. Jones, E. C. Thomas, J. -P. Villian, J. -C. Cerisier, C. Senior, C. Hanuise, R. D. Hunsucker, G. Sofko, J. Koehler, E. Nielsen, R. Pellinen, A. D. M. Walker, N. Sato and H. Yamagishi, DARN/SuperDARN: a global view of the dynamics of high-latitude convection, *Space Sci. Rev.*, 71, 761-796, 1995.

Kawano, H. and C. T. Russell, Survey of flux transfer events observed with the ISEE-1 spacecraft; dependence on the interplanetary magnetic field, *J. Geophys. Res.*, 102, 11307-11313, 1997.

Lockwood, M. and C. J. Davis, On the longitudinal extent of magnetopause reconnection pulses, *Ann. Geophys.*, 15, 865-878, 1996.

Luhmann, J. G., R. J. Walker, C. T. Russell. N. U. Crooker, J. R. Spreiter and S. S. Stahara, Patterns of potential magnetic field merging site on the dayside magnetopause, *J. Geophys. Res.*, 89, 1739-1742, 1984.

Maynard, N. C., W. J. Burke, P. E. Sandholt, J. Moen, D. M. Ober, M. Lester, D. R. Weimer and A. Egeland, Observations of simultaneous effects of merging in both hemispheres, *J. Geophys. Res.*, 106, 24551-24577, 2001.

McWilliams, K. A., T. K. Yeoman and G. Provan, A statistical survey of dayside pulsed ionospheric flows as seen by the CUTLASS Finland HF radar, *Ann. Geophys.*, 18, 445-453, 2000.

Newell, P. T.,
latitude of tl
Geophys. Re
Newell, P. T.,
interplanetar
into the m
magnetopau:
1995.
Onsager et al. ;
Øieroset, M.,
Energetic ic
relationship
activity, J. A
Pinnock, M., A
High spatial
Ann. Geophy
Pinnock, M., A
Conjugate F
merging: A C
443-454, 199
Provan, G., T.
the IMF By c
dayside ionos
Lett., 26, 521-
Rodger, A. S.
boundaries, A
Rodger, A. S.,
on transient
magnetopause
Safrankova, J., Z

Mc
tec
the
eta
cra
une
wei
froi
sur:
spe
firs
Sou
a si
Sva
Hen
/
lites
the
topa
in m
tegri
The
extei
on tl
it is
whei
the !
with
layer
magr
plane
of me
plyin;
Sisco
the I!
the in
Fii
of an
tion r
the si
occur
They
mergii
magne
All
the do
itudes
magne
per is
cations

Maynard et al. [1995] showed that cusp potential patterns are well ordered in the inertial reference frame in which the throat between the two cells resides near noon for both polarities of B_Y, in agreement with the statistical position of cusp particle precipitation. The convection pattern shown in the lower left quadrant of Plate 1 is placed on a magnetic latitude versus magnetic local time (MLT) grid. It was derived using the *Weimer* [2001] (W2K) convection model, represented in inertial coordinates. In corotating coordinates the split between the two cells is enclosed within the large afternoon cell (cf. Figure 2 of *Maynard et al.* [1995]). Patterns derived from SuperDARN radar measurements accumulated during the rocket flights at 2-min intervals qualitatively match the corotating version of the illustrated W2K pattern. The red circle schematically represents the coverage of the all-sky imager during the rocket flight. Within the circle we indicate the meridian scan direction, as well as the trajectories of the rocket and the Defense Meteorological Satellite Program (DMSP) F13 satellite. At the time of the flight, the all-sky image and a meridian scanning photometer scan line included segments of both the morning and afternoon convection cells. Estimated projections of the Northern and Southern Hemisphere merging lines are represented by the orange and yellow dashed lines, respectively. They are located on two sides of the gap between the afternoon and morning convection cells. With antiparallel merging, the lengths of field lines connecting the ionosphere to Northern and Southern Hemisphere merging sites are quite different. Convection in the morning cell may continue in the sunward and eastward directions for several minutes after field lines first become open. This reflects Alfvén transit times required for signals generated at the Southern Hemisphere merging site (top left plot) to reach the northern ionosphere and turn plasma flow poleward and westward.

The bottom right plot of Plate 1 shows similarly color-keyed regions superposed on a 630.0 nm all-sky image. Note that the image is rotated 180° from its representation on the potential pattern. This gives it the same orientation used in presenting all-sky images below. *Sandholt et al.,* [1998] have categorized cusp aurora according to the location and perceived source. During southward IMF, the cusp aurora (Type 1) has a sharp lower border, and forms move poleward. During northward IMF, the poleward border of the cusp aurora becomes sharper as merging occurs on the poleward edge of the cusp. They referred to this as Type 2. Along the orange locus for Northern Hemisphere merging, Type 1 aurora would occur on the lower part of the curve. As the IMF turns northward the active merging site would move up to the poleward portion of the curve, creating Type 2 aurora.

2.1. IMF B_Y spatially bifurcates the cusp

The scenario outlined in Plate 1 assumes that with a significant IMF B_Y, merging occurs at high magnetic latitudes and implies that the cusp bifurcates into separate regions connected to Northern and Southern Hemisphere merging sites. Plate 2 expands on this concept [*Maynard et al.,* 2001c]. Plates 2a and 2b show all-sky images taken at 630.0 nm during the rocket flight. The positions of the rocket and DMSP F13 are represented by red and green dots, respectively. Plate 2c shows the trace of E_{KL} calculated from Wind data with the determined best lag [*Maynard et al.,* 2001c]. The peaks labeled 1–8 are those that were correlated with the rocket (Figure 1) and optical data. The shorter than normal-advection lag time restricts this interaction to Southern Hemisphere merging. Considering the rocket correlation, the optical patterns, and data from a DMSP overpass, *Maynard et al.* [2001c] placed an approximate boundary, shown by the white hook, to separate emissions from Southern Hemisphere merging (to the left of and inside the hook) from those associated with Northern Hemisphere merging. Plates 2d-2i show all-sky images of 557.7 emissions, taken approximately every 10 s, spanning the period of peaks 1 and 2. The white hook separates the same area as noted in Plates 2a and b. The arrows point to the region at the base of the hook responding to changes in E_{KL}. Note the increase in intensity at the bottom edge of the hook in Plate 2e and lesser emissions 10 s before and after. The 557.7 nm emissions are equatorward of the major cusp particle precipitation located by the 630.0 nm arc near zenith in Plate 2. Similarly *Rodger and Pinnock* [1997] found that the first responses of flux transfer events are velocity increases at the open/closed boundary 100 to 200 km equatorward of the red-arc cusp. The broader response of peak 2 reflects the corresponding longer E_{KL} peak. In response to each of the eight peaks, temporal peaks in 557.7 nm emissions may be found within the hook that are similar to those shown in Plates 2d-2f (compare Figure 1). As shown below, emissions from the region eastward of the hook must respond differently. Bifurcation effects can also be seen in the snapshot provided during the flight of DMSP F13 over Svalbard (see Plate 11 of *Maynard et al.* [2001c]).

Spatial bifurcation of cusp source regions is a necessary consequence of high-latitude merging. The antiparallel merging criterion indicates that this should occur whenever B_Y is the dominant IMF component. The merging site moves poleward of the cusp when IMF B_Z is northward and dominant. For a purely southward IMF B_Z, merging proceeds in the sub-solar region. The range of IMF orientations for which merging transitions from near the sub-solar magnetopause to high latitudes cannot be specified from observations discussed here.

2.2. IMF B_X influence on the timing of merging

B usually lies in or near the plane of constant IEF phase. Increasing B_X increases the tilt of the phase plane. *Maynard et al.* [2001c] found that the aurora to the east of the hook in Plate 2a could be harmonized with E_{KL} measurements if a longer lag time was used. The lag time for Northern Hemisphere merging was determined from intervals when IMF B_Z briefly turned northward. These brief intervals of merging on the poleward edge of the cusp provided Type 2 optical signatures [*Sandholt et al.*, 1998] which occurred just before DMSP entered the cusp region. Since DMSP did not detect sunward convection in the cusp, we conclude that the large-scale pattern did not change during the brief northward turnings.

The difference between the Southern and Northern Hemisphere lag times can be interpreted qualitatively as a result of the tilt of the phase plane. The tilt angle varied with the strength of B_X. Thus, with a significant B_X and B_Y larger than B_Z, the two hemispheres respond to the same elements of the solar wind stream at significantly different times. The degree of dominance of B_X controls the temporal duration of bifurcation. Note that temporal bifurcation such as that observed by *Maynard et al.* [2001c] disappears when sub-solar merging is dominant as would be the case for negative B_Z dominated conditions.

3. CONVECTION RESPONSE

It is important to note that in many areas the *Heppner and Maynard* [1987] convection patterns for B_Y positive and negative are not conjugate. Yet B_Y positive or negative patterns must coexist simultaneously in the two hemispheres. MHD simulation results reported by *Maynard et al.* [2001a,b] provide further insight into the dynamic processes governing ionospheric convection. The connection between the sash and the plasma sheet along the magnetotail flank varies temporally and spatially. In the simulation, closed field lines threading through the low-magnetic field connection region map to different potentials in the two ionospheres. One implication is that time-varying structures in the sash-plasma sheet connection region serve to decouple the two ionospheres, allowing different convection patterns to coexist [*Maynard et al.*, 2001a].

A simulated switch in the polarity of IMF B_Y [*Maynard et al.*, 2001b] reproduced experimentally observed effects reported by *Ridley et al.*, [1997, 1998]. Namely, the main convection pattern does not appear to change until ~8 min after first contact of the altered IMF with the magnetosphere. However, the change starts over the entire polar cap. In the simulation, the convection pattern change did not begin until the cusp-mantle currents reversed both in the vicinity of the sub-solar magnetopause and above the cusp. The simulated

effects of merging with the new polarity were discernible ~3 min after first contact. For a short time, merging can proceed at a new site without an observable change in large-scale convection patterns.

Both simulated and experimental results [*Maynard et al.*, 2001b,c] imply that under B_Y-dominated conditions the small crescent-shaped convection cell is driven by merging in the opposite hemisphere. This is a direct consequence of spatial bifurcation of the cusp. For B_Y positive, merging in the Southern Hemisphere produces open field lines that connect to the Northern Hemisphere cusp and drape over the nose of the magnetosphere. They are dragged back along the dawn flank as the magnetosheath plasma, in which they are embedded, propagates downtail. A similar set of open field lines exists on the dusk side that maps to the southern ionosphere. *Maynard et al.* [2001a] showed that these open field lines form thick boundary layers along the flanks of the magnetotail near the equatorial plane and are part of the small convection cell. This open flux reconnects in the sash and subsequently returns to the dayside as closed field lines that complete the small-cell circulation cycle. This suggests that a significant fraction of lobe flux reconnects on the flanks rather than in the center of the magnetotail.

4. BOUNDARY LAYER IMPLICATIONS OF HIGH-LATITUDE MERGING

Having determined that boundary layers on the flanks of the magnetotail may be topologically open and spatially thick, we next examine some implications of high-latitude merging on boundary layer properties around the entire magnetopause.

Plate 3 shows the first open field lines attached to the northern (blue) and southern (red) ionospheres as the switch in the polarity of B_Y reached the magnetosphere [from *Maynard et al.*, 2001b]. Plate 3a represents the old $B_Y > 0$ ($\theta = 90°$) orientation with blue (Northern Hemisphere) field lines draped over the pre-noon magnetopause. Conversely, red (Southern Hemisphere) field lines drape over the post-noon magnetopause. This layer is thin near the nose and thickens along the flanks. Bifurcation of the cusp from high-latitude merging has the consequence of generating open boundary layers. *Maynard et al.* [2001a] noted that a thick, open boundary layer shields closed field line regions from viscous forces exerted through magnetosheath plasma flow. Magnetic tension provides the dominant tailward force on the open field lines.

The next three plots of Plate 3 show the evolution of the open boundary layer as IMF B_Y changed polarity. Plate 3b shows that two minutes later only closed (black) field lines are visible in the sub-solar region. Four minutes after the reversal contacts the magnetopause (Plate 3c), the first

Elphinstone R.D., D.Hearn, J.S.Murphree, and L.L.Cogger, Mapping using the Tsyganenko long magnetospheric model and its relationship to Viking auroral images, *J. Geophys. Res.*, 96, 1467-1480, 1991.

Erlandson, R.E., L.J.Zanetti, T.A.Potemra, P.F.Bythrow, and R.Lundin, IMF B_Y dependence of region 1 Birkeland currents near noon, *J. Geophys. Res.*, 93, 9804-9814, 1988.

Frank-Kamenetsky, D.A., *Lectures on plasma physics*, Atomizdat, Moscow, 286, 1968 (in Russian).

Haerendel G., G.Paschmann, N. Scopke, et al., The frontside boundary layer of the magnetosphere and the problem of reconnection, *J. Geophys. Res.*, 83, 3195-3216, 1978.

Heikkila, W.J., Magnetic reconnection, merging, and viscous interaction in the magnetosphere, *Space Sci., Rev.*, 53, 1, 1990.

Hones E.W., J.R.Asbridge, S.J.Bame, et al., Measurements of magnetotail plasma flow with Vela 4B, *J. Geophys. Res.*, 77, 5503-5521, 1972.

IijimaT., and T.A.Potemra, The amplitude distribution of field-aligned currents at northern high latitudes observed by TRIAD, *J. Geophys. Res.*, 81, 2165-2174, 1976.

Jaggi, R.K., and R.A.Wolf, Self-consistent calculation of the motion of a sheet of ions in the magnetosphere, *J . Geophys., Res.*, 78, 2842-2852, 1973.

Newell P.T., and C.-I. Meng, Mapping the dayside ionosphere to the magnetosphere according to particle precipitation characteristics, *Geophys. Res. Letters*, 19, 609-612, 1992.

Newell P.T., W.J.Burke, E.R.Sanchez, C.-I.Meng, M.E.Greenspan, and C.R.Clauer, The low-latitude boundary layer and the boundary plasma sheet at low altitude: prenoon precipitation regions and convection reversal boundaries, *J. Geophys. Res.*, 96, 21013-21023, 1991.

Ohtani S., T.A.Potemra, P.T.Newell, L.J.Zanetti, T.Iijima, M. Watanabe, M.Yamauchi, R.Elphinstone, O.de la Beaujardiere, and L.G.Blomberg, Simultaneous prenoon and postnoon observations of three field-aligned current systems from Viking and DMSP-F7, *J. Geophys. Res.*, 100, 119-136, 1995.

Sato, T., Possible sources of field-aligned currents, *Rep. Ionos. Space Res. Japan*, 28, 179-186, 1974.

Shiokawa, K., W.Baumjohann, and G.Haerendel, Braking of high-speed flows in the near-Earth tail, *Geophys. Res. Lett.*, 24, 1179-1182, 1997.

Sonnerup, B.U.O. Theory of the low-latitude boundary layer, *J. Geophys. Res.*, 85, 2017-2026, 1980.

Tsuda, T., Effective viscosity of a streaming collision-free plasma in a weakly turbulent magnetic field, *J. Geophys. Res.*, 72, 6013-6020, 1967.

Vasyliunas, V.M., Mathematical models of magnetospheric convection and its coupling to the ionosphere, *in Particles and fields in the magnetosphere*, Dordrecht, 1022-1032, 1970.

Wolf, R.A., Calculations of magnetospheric electric fields, *in Magnetospheric Physics*, ed. By B.M. McCormac, 167-177, D.Reidel, Hingham, Mass., 1974.

Yamauchi, M., R.Lundin, and J.Woch, The interplanetary magnetic B_Y effects on large-scale field-aligned currents near local noon: contributions from cusp part and noncusp part, *J. Geophys. Res.*, 98, 5761-5767, 1993.

Yang, Y.S., R.W.Spiro, and R.A.Wolf., Generation of region 1 current by magnetospheric pressure gradients, *J. Geophys. Res.*, 99, 223-234, 1994.

O. A. Troshichev, Department of Geophysics, Arctic and Antarctic Research Institute, 38 Bering str., 199397, St.Petersburg, Russia, olegtro@aari.nw.ru

The Aurora as Monitor of
Solar Wind-Magnetosphere Interactions

Per Even Sandholt

Department of Physics, University of Oslo, Norway

Charles J. Farrugia

Space Science Center, University of New Hampshire, Durham, New Hampshire

We review recent work on dayside auroral responses to IMF states and transitions. Four classes of typical dayside auroral transients are isolated and all can be explained as signatures of electron precipitation from different modes/configurations of magnetopause reconnection. We supplement the ground-based optical auroral observations with radar data on ionospheric convection and particle precipitation obtained from satellites in polar orbit. Class 1 corresponds to southward IMF and is characterized by auroral equatorward boundary intensifications (EBIs) on open LLBL field lines, where low-altitude precipitation shows an electron edge and a clear low-energy cutoff in the ions. EBIs are followed by poleward moving auroral forms (PMAFs) in the region of cusp/mantle precipitation. PMAFs have been observed in conjunction with a staircase ion energy-latitude dispersion, supporting their reconnection interpretation. Class 1 observations result from pulsed reconnection at sub-cusp latitudes. In class 2, corresponding to northward IMF B_z with a significant B_y component, the aurora is characterized by east-west expanding forms which are accompanied by activations of lobe cell convection. In our view, class 2 represents the auroral response to bursts of high-latitude reconnection poleward of one cusp. For strongly northward IMF, we have class 3, which extends class 2 and constitutes the signature of high-latitude reconnection occurring near-simultaneously poleward of both cusps. The contracting poleward boundary of the cusp aurora is subject to sequential brightenings/poleward expansions at times of "reverse" 2-cell convection. These observations may be explained by the addition of closed flux to the dayside when the magnetosphere captures magnetosheath flux tubes. Finally, class 4 auroral activity pertains to a

Earth's Low-Latitude Boundary Layer
Geophysical Monograph 133
Copyright 2003 by the American Geophysical Union
10.1029/133GM34

Observations of Postnoon Auroral Bright Spots With High Temporal and Spatial Resolution at Zhongshan Station, Antarctica

Yong Ai

Department of Optoelectronic information engineering, Wuhan University, Wuhan, China

Huigen Yang

Polar Research Institute of China, Shanghai, China

M Kikuchi

National Institute of Polar Research, Tokyo, Japan

By analyzing ground-based optical observations at Zhongshan, Antarctica, several results can be obtained. Most of the bright spots have a vortex structure, and they are rather transient features lasting typically for only a few minutes (some last less than 30 s). The typical size of the spots is about 100-300 km. The intensity of the bright spots of 557.7 nm emissions can surpass 13 kR. The occurrence of the bright spots seems rather frequent and favors low magnetic activity. Usually the bright spots are accompanied by multi-band arcs and most arcs move poleward. The latitudinal extent of the arcs can surpass 300 km, the intensification of the spot is mainly caused by an enhanced middle level energy (3~10 keV) flux, the occurrence height of the bright spot is about 150 ~200 km. Ground-based observations confirms the existence of auroral intensification between 1400 and 1700 MLT. Magnetic pulsations occurred in conjunction with bright spots. The pulsations had periods of 41, 20.8, 10.4 and 5.2 minutes, with the short periods corresponding to those of the PMAFS in the noon sector.

1. INTRODUCTION

Auroral phenomena occur at all local times, but their characteristics vary substantially from the dayside to the night side. The postnoon region seems to have a special significance in magnetospheric physics. The region I upward field-aligned current maximizes in the 1400-1600 MLT sector, and measurements have shown strong particle precipitation occurring there. These relate to a general enhancement of auroral intensity in this region.

Earth's Low-Latitude Boundary Layer
Geophysical Monograph 133
Copyright 2003 by the American Geophysical Union
10.1029/133GM37

Using the ISIS 2 particle detector, *McDiarmid et al.* [19 75] constructed average intensity contour maps for 150 eV electrons, which showed a maximum near 1500 MLT and 75° invariant latitude. *Evans* [1984] also found a statistical maximum in energy flux in the 1300-1500 MLT sector at invariant latitudes near 78°, and determined that the energy flux was due to precipitating low-energy (<3 keV) electrons. *Bruning et al.* [1990] presented Viking in situ measurements above the acceleration region corresponding to a large 1400 MLT auroral bright spot. The source plasma parameters were typical of those of magnetosheath plasma; the field-aligned potential drop deduced from the upward ion beams was in the range of 1-3 kV.

Newell et al. [1996] surveyed electron acceleration events in precipitation data taken by the DMSP satellites over a 9-year interval. They reported that the "midday gap" is real, with noon easily the least likely local time for observing electron acceleration; the "14 MLT (magnetic local time) hot spot" is also real, although it is centered about 15 MLT and is distinct from the rest of the oval only for northward interplanetary magnetic field conditions; A weaker "warm" spot exists in the 6-9 MLT region. *Lui et al.* [1987] argued that the dayside sector is a dynamic region with activity comparable to the night side region even for disturbed periods.

The low-latitude boundary layer with a strong plasma shear is proposed as the region of viscous interaction between the magnetosheath and the magnetosphere [*Eastman et al.*, 1976]. On the other hand, field line merging has been suggested as the main energy and momentum transfer process. Measurements from the ISEE and Prognoz 7 satellites indicate that the interface region between the magnetosheath and the magnetosphere is highly structured and variable [*Lundin and Dubinin*, 1984]. Therefore, the ionospheric projection of this region cannot be a simple one, and the auroral display in the dayside may reveal clues about the complicated structure of the magnetopause and the polar cusp region.

Compared with satellite observations, the ground-based aurora observation has many advantages in providing high temporal and spatial resolution data. Postnoon auroral bright spots are observed at Zhongshan station, Antarctica during winter. The station is located at geographic 69.4° S in latitude and 76.4° E in longitude or 74.5° (L=13.9) in magnetic latitude. Magnetic local time (MLT) is about 1.3 hours ahead of universal time (UT), that is MLT = UT+1.3, and local time (LT) is about 3.8 hours ahead of MLT, that is LT = MLT+3.8. During May to August, the station can make observation from 1600 LT to 0700 LT (about 1200 MLT to 0300 MLT). The auroral oval is just over the station around 1400 MLT (about 1800 LT), so the station is a perfect place for postnoon auroral studies. For optical measurements of aurora, the station is equipped with an all-sky TV camera, an all-sky CCD camera, a high speed multi-channel meridian scanning photometer (MSP), and magnetometers.

Although the postnoon auroral bright spots have been studied by many scientists with satellite data, reports from ground-based optical observations are still very few. The purpose of the present paper is to provide ground-based auroral bright spots observations with a high temporal and spatial resolution.

2. OBSERVATIONS

The settings of the parameters for MSP are as follows: scanning period: 4s; sampling frequency: 100 points/s; scanning angle: 0-180 deg.; detecting wavelengths: 630.0 nm, 557.7 nm, 427.8 nm. Thus the MSP provides excellent mappings of the small-scale auroral structures along the magnetic meridian. In processing MSP data, we assume the emission height of 557.7 nm (427.8 nm) and 630.0 nm are 110 km and 250 km, respectively. Normally, the observed emissions should be adjusted to the atmospheric absorption effect and the Van Rijhn effect, an effect caused by uneven optical thickness of the source when looking at different angles. Because the atmospheric absorption effect and the Van Rijhn effect counteract each other, and local parameters of atmospheric absorption is difficult to evaluate, we are not going to consider the atmospheric absorption effect and the Van Rijhn effect in this paper and this will not cause apparent errors, especially in the case of analyzing ratios from a few different emission lines.

The exposure time of all-sky CCD camera is set at 10 s for 557.7 nm emission , every 15 s one image of the aurora is taken (630.0 nm emission also can be observed by changing the 630.0 nm filter). All-sky TV camera system comp- rises a HARPICOK (High-gain Avalanche Rushing Amorphous Photoconductor) TV camera fitted with a fish-eye lens. The lowest pickup light level is ~0.02 lx at a video rate of 30 frames, which is about one order of magnitude lower than that of a SIT (Silicon Intensifier Target) tube.

Figure 1 gives an 8 hour overview of the photometer data set for 557.7 nm emissions on July 11, 1999, that is from 1045 to 1845 UT (about 1200 to 2000 MLT). The measurements are displayed as stacked plots of 557.7 nm intensities as functions of the photometer scan angle along Zhongshan magnetic meridian. The X-axis indicates the universal time and Y-axis shows the intensities distribution of the aurora along the magnetic meridian. From Figure 1, we can see clearly the strong dayside auroral intensification between 1400 and 1700 MLT. Figure 2 (a) plots one-hour photometer data from 1446 to 1546 UT for 557.7 nm emissions on the same day at that of Figure 1, so we can see the bright spots more clearly. The brightest spot occurred at about 1457 UT, and the intensity of this spot was over 13 kR. In order to see this bright spot more clearly, we enlarged this area in Figure 2(b).

Figure.3 shows the all-sky photos of the bright spots taken by the all-sky TV camera on July 11, 1999 at about 1457 UT. Note that the top is south, the bottom is north,

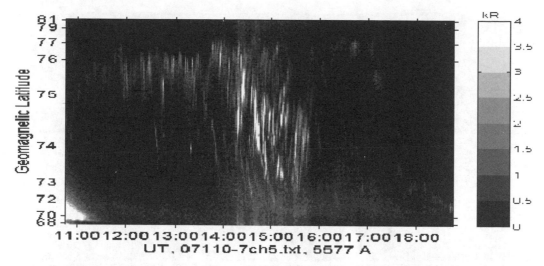

Figure 1. Keogram of the optical aurora (557.7 nm) observed at Zhongshan Station between 1045 and 1845 UT on July 11, 1999.

the right is east, and the left is west. From Figure 3, we can see a new bright spot with ray structure developing just over the Zhongshan station at 14:56:44 UT, the luminosity increased dramatically, then the spot with a curl shape moved poleward and east ward very fast. After 3 minutes, the spot disappeared.

Precipitating electrons of different energies interact differently with atmospheric neutrals to produce diverse auroral emission lines. A qualitative appreciation for variations of the average energy of precipitating electrons, based on optical measurements, can be obtained from the ratios of observed auroral emission intensities at different wavelengths [*Egeland A, et al.*, 1994]. Here, we consider the bright spot in Figure 2 (b). Figure 4 (a) plots, as a functions of universal time, the intensities obtained by integrating emissions from latitude 74 S to 75 S (this bright spot is just over the Zhongshan station and the

atmospheric absorption effect and the Van Rijhn effect are minimum) for 557.7 nm, 630.0 nm and 427.8 nm emission lines respectively, which qualitatively indicates the total fluxes of precipitating particles along the designated latitudinal extent above Zhongshan station during the occurrence of bright spot. Figure 4 (b) plots the evolution of integrated intensity (from $5°$ to $175°$ elevation angles) of postnoon auroral emissions taken by meridian scanning photometer at Zhongshan Station between 1045 and 1845 UT on the July 11, 1999. From the Figure, we can clearly see a postnoon auroral intensification between 1400 and 1600 UT. Figure 5 gives the results of integrating intensity ratio of I(4278)/I(6300) (a) and I(4278)/I(55 77) (b). The results qualitatively indicate changes in the average energy of the precipitating electrons along the meridianal scan above Zhongshan station.. Figure 6 presents the

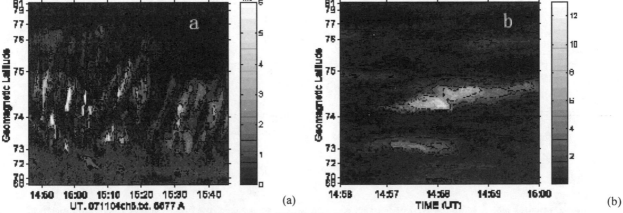

Figure 2. Keogram of the optical aurora (557.7 nm) observed at Zhongshan Station on July 11, 1999.
(a) Between 1445 and 1545 UT, (b) Between 1456 and 1500 UT.

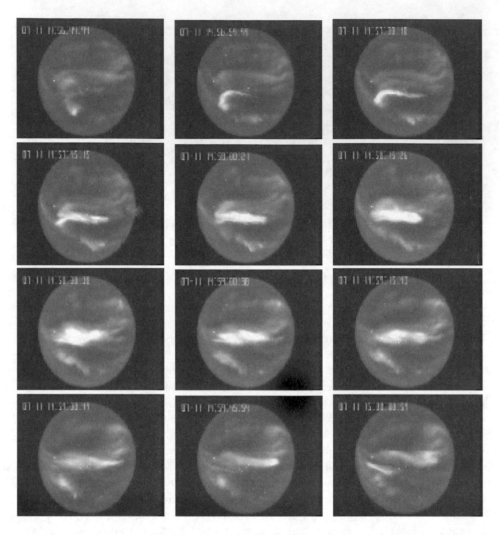

Figure 3. Evolvement of aurora bright sports taken by all-sky TV camera at Zhongshan Station on July 11, 1999

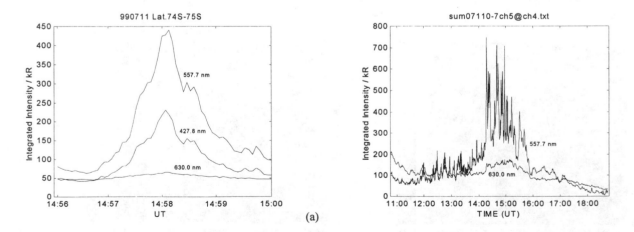

Figure 4. Evolvement of integrated intensity of a bright sport event taken by meridian scanning photometer at Zhongshan Station on July 11, 1999. (a) Between 1456 and 1500 UT. (b) Between 1045 and 1845 UT.

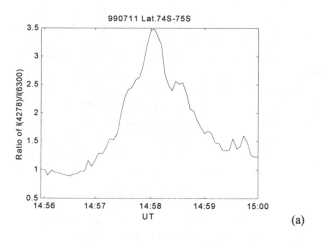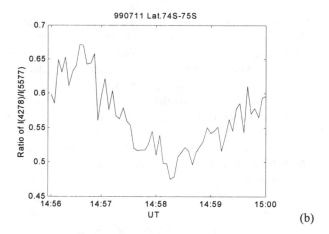

(a) (b)

Figure 5. Evolvement of ratio of I(5577)/ I(6300) (a) and I(4278)/ I(5577) (b) taken by meridian scanning photometer at Zhongshan Station between 1456 and 1500 UT on July 11, 1999

change of the H component of the local magnetic field between 1400 and 1600 UT on the July 11,1999, and also shows the wave spectrum of magnetic pulsation. Table 1 gives three-hourly averaged Kp indices on the day.

3. DISCUSSION

The dayside auroral region is linked through magnetic field lines to the frontside magnetosphere where solar wind plasma can penetrate through the dayside magnetopause or enter via the polar cleft or cusp. Investigations of auroral phenomena in the dayside region may thus provide   insights regarding the dynamic processes in the dayside magnetosphere.

Figure 1 shows that the dayside auroral oval was moving equatorward during the whole afternoon, and there is a strong luminosity intensification zone exists between 1300 and 1600 UT (about 1400 to 1700 MLT). From Figure 1 to Figure 4, we can see that the luminosity intensifications were accompanied by the occurrence of bright spots and periodic poleward motion of the aurora arcs. The PMAF (poleward moving auroral forms) may be considered as ionospheric signatures of flux transfer events (FTEs). The intensity of the bright spots can surpass 13 kR, which can be comparable with that of nightside discrete arcs. The movements and the shapes of the spots can change very fast. From several years of observations, we found that most of the postnoon bright spots have a vortical structure and last only a few minutes. The typical size of the bright spots is around 100~300 km.

From Figure 3 and Figure 4, we can see the evolution of the intensity and latitude position of an enlarged aurora bright spot. Figure 4 (a) shows that all three emissions of 557.7 nm, 630.0 nm and 427.8 nm increased

as the occurrence of bright spot, indicating that the total energy flux and high energy (3~10 keV) fluxes reached a maximum and then decreased during the bright spot process. So the appearance of the bright spots must be accompanied by electron acceleration. Figure 4 (b) confirms the existence of an auroral intensification between 1400 and 1600 UT. Figure 5 (a) shows that the ratio of I(4278)/I(6300) has a positive peak distribution, indicating qualitatively that the average precipitating electron energies along the meridianal scan above Zhongshan station increased as the bright spot appeared. And Figure 5 (b) shows that the ratio of I(4278)/I(5577) has a negative peak distribution during the occurrence of the

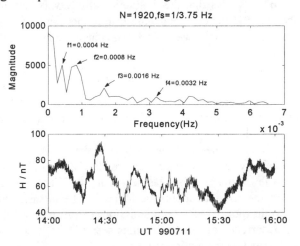

Figure 6. The change of H component of the local magnetic field between 1400 and 1600 UT on the July 11,1999 (down panel) and wave spectrum of magnetic pulsation (top panel).

Table 1	Kp Three-hourly Indices	(July 11, 1999)						
Time(UT)	1	2	3	4	5	6	7	8
Kp	0+	1-	1	1-	1	1-	1	2+

bright spot. This result is more meaningful. The result illustrates that the increase of the average precipitating electron energies of 557.7 nm emission is larger than that of 427.8 nm emission. As there is no quenching effect, the energy of the precipitating particles should be less than 10~20 keV, the intensification of the spot mainly being caused by the increasing of high energy (3~10 keV) flux, because the 5577 emission is sensitive to electrons of a few keV or higher, and 4278 emission depends more on the total energy flux. The occurrence height of the bright spot may be about 150~200 km. Usually the bright spots are accompanied with the occurrence of multi-band arcs and the most motions of the arcs are poleward, latitudinal extent of the arcs can surpass 300 km.

Figure 6 shows that magnetic pulsations corresponded with bright spots, and main spectrum are 0.0004, 0.0008 , 0.0016 and 0.0032 Hz. So the main pulsation periods include 41, 20.8, 10.4 and 5.2 minutes. The period of 5.2 minutes coincides with the PMAF occurrence period in Figure 2. Magnetic pulsations like these were also observed by *Rostoker et al.* [1992]. The pulsation period could be the characteristic period of a wave acting on the magnetopause / LLBL region, probably in the flank region.

As to the source region of 1500 MLT auroral bright spots, *Liou et al.* [1999] use the auroral images from the Polar ultraviolet imager (UVI) and simultaneous particle observation from DMSP and found precipitating particle may come from plasma sheet, the low-latitude boundary layer, and the plasma mantle.

Table 1 shows that the magnetic activity was low on July 11, 1999, even though the bright spot activity was very strong . And we found this phenomena is very common.

4. CONCLUSIONS

By this paper we can get some new results from the ground-based observation of postnoon auroral bright spots as follows.

(1) Most of the bright spots have a vortical structure, and they are rather transient features lasting typically for only a few minutes (some last less than 30 s). The typical size of the spots is about 100-300 km. The intensity of the bright spots of 557.7 nm emissions can surpass 13 kR. The occurrence of the bright spots seems rather frequent and favors low magnetic activity.

(2) Usually the bright spots are accompanied with the occurrence of multi-band arcs and most arcs move poleward. The latitudinal extent of the arcs can surpass 300 km. The intensification of the spots is mainly caused by the enhanced middle level energy (3~10 keV) fluxes, the occurrence height of the bright spot is about 150 ~200 km.

(3) Ground-based observations confirms the existence of auroral intensification between 1400 and 1700 MLT. The phenomena of the magnetic pulsation during the occurrence of bright spots were observed, the main pulsation periods include 41, 20.8, 10.4 and 5.2 minutes. The period of 5.2 minutes coincide with the PMAF occurrence period in postnoon sector. The number of pulsation period could be the characteristic period of a wave acting on the magnetopause / LLBL region, probably in the flank region.

REFERENCES

Bruning, K., L. P. Block, G. T. Marklund, L. Eliasson, *et al*, Viki- ng observations above a postnoon aurora, *J. Geophys. Res.*, 95, 6039, 1990.

Eastman, T. E., E. W. Hones, Jr., S. J. Bame, and J. R. Asbridge, The magnetospheric boundary layer, site of plasma, mimentum, and energy transfer from the magnetosheath into the magnetos- phere, *Geophys. Res. Lett.*, 3, 685, 1976.

Egeland, A., W. J. Burke, N. C. Maynard, *et al.*, Ground and sate- llite observations of postdawn aurorae near the time of a sudden storm commencement, *J Geophys Res.*, 99, 2095, 1994.

Evans, D. S., The characteristics of a persistent auroral arc at high latitude in the 1400 MLT sector. *In The Polar Cusp*, edited by J A Holter and A Egeland, p.99, Hingham, Mass, 1984.

Liou K., P. T. Newell, C.-I. Meng, and T. Sotirelis, Source region of 1500 MLT auroral bright spots: Simultaneous Polar UV- images and DMSP particle data, *J. Geophys. Res.*, 104, 24587, 1999.

Lui A. T. Y., D. Venkatesan, G. Rostoker, C. Murphree, D. Anger , L. Cogger and T. A. Potemra, Dayside auroral intensifications during an auroral substorm. *Geophys. Res. Lett.*, 14, 415, 1987.

Lundin, R., and E. Dubinin, Solar wind energy transfer regions inside the dayside magnetopause, I, Evidence for magnetosh- eath plasma penetration, *Planet. Space Sci.*, 32, 745, 1984.

McDiarmid, I. B., J. R. Burrows, and E. E. Budzinski, Average characteristics of magnetospheric electrons (150 eV to 200 eV) at 1400 km, *J. Geophys. Res.*, 80, 73, 1975.

Newell, P. T., K. M. Lyons, and Ching-I. Meng, A large survey of electron acceleration events, *J. Geophys. Res.*, 101, 2599, 1996.

Rostoker, G. A., B. Jackel, and R. L. Arnoldy, The relationship of periodic structures in auroral luminosity in the afternoon sector to ULF pulsations, *Geophys. Res. Lett.*, 19, 613, 1992.

Yong Ai, Department of Optoelectronic information engin- eering, Wuhan University, Wuhan 430072, China. (e-mail: aiyong@ public.wh.hb.cn)

Huigen Yang, Polar Research Institute of China, Shanghai 200129, China. (e-mail: huigen@mail2.online.sh.cn)

M Kikuchi, National Institute of Polar Research, Tokyo 173- 8515, Japan. (e-mail: kikuchi@nipr.ac.jp)

Local Boundary Layer Properties from Non-Local Processes Illustrated by MHD Simulations

G. L. Siscoe

Center for Space Physics, Boston University, Boston, Massachusetts

K. D. Siebert

Mission Research Corporation, Nashua, NH

Boundary layers associated with the magnetopause of the magnetosphere illustrate a general property of magnetized cosmic plasmas to exhibit local properties that arise from non-local processes. Here we develop this theme with examples illustrated by images obtained through global MHD simulation, perhaps the best method currently available to explore global connections of magnetospheric domains. Among local properties governed by non-local processes thus illustrated are the following: 1. a magnetically closed boundary layer next to a magnetically open boundary layer with no discontinuity between them in density or velocity; 2. boundary layer plasma, both open and closed, at equatorial latitudes resulting from magnetic reconnection at high latitudes but not involving flow from high latitudes to equatorial latitudes; 3. closed boundary layer plasma, thus produced, feeding relatively cold and dense plasma to the region of the tail normally occupied by the hot, thin plasma sheet; and 4. strong dependence on the pointing direction of the interplanetary magnetic field of volumes occupied by boundary layers. Although not all of these concepts are new, this paper presents images crafted from MHD simulations to illustrate them.

1. ILLUSTRATING BOUNDARY LAYER PROPERTIES WITH MHD SIMULSATIONS

To illustrate examples of local boundary layer properties determined by non-local processes, we use results obtained with a global magnetospheric MHD numerical code, the Integrated Space Weather Prediction Model (ISM). Properties of the ISM code have been described elsewhere, most fully in *White et al.* [2001]. Basically, the code integrates the standard MHD equations over a volume that extends from 40 R_e upwind from Earth to 300 R_e downwind from Earth, and 60 R_e radially from an axis through Earth parallel to the solar wind flow direction. It handles connection to the ionosphere in a way that is unique among global MHD codes in that it uses the same set of equations from the base of the ionosphere to the outer boundaries of the grid, just described. The single set of equations that it uses automatically becomes the appropriate continuum mechanics set for each domain from the ionosphere to the solar wind and segues smoothly between domains.

One advantage of using global MHD simulations to address non-local processes is that they integrate Newton's equations and Maxwell's equations simultaneously and self consistently everywhere. Thus the dynamic and thermal pressure fields and the magnetic fields are guaranteed

Earth's Low-Latitude Boundary Layer
Geophysical Monograph 133
Copyright 2003 by the American Geophysical Union
10.1029/133GM38